Historical & Cultural Astronomy

Historical & Cultural Astronomy

More information about this series at http://www.springer.com/series/15156

Alexus McLeod

Astronomy in the Ancient World

Early and Modern Views on Celestial Events

Springer

Alexus McLeod
University of Connecticut
Storrs, CT
USA

ISSN 2509-310X ISSN 2509-3118 (electronic)
Historical & Cultural Astronomy
ISBN 978-3-319-23599-8 ISBN 978-3-319-23600-1 (eBook)
DOI 10.1007/978-3-319-23600-1

Library of Congress Control Number: 2016941290

Cover illustration: The series cover image, "Astronomer by Candlelight," was painted by Gerrit Dou in the style of the school of Leiden "fijnschilders;" Dou was born in Leiden in 1613 and died there in 1675.

Image Credit: Jarek Tuszynski/CC-BY-SA & GDFL.

Printed on acid-free paper

This Springer imprint is published by Springer Nature
The registered company is Springer International Publishing AG Switzerland

Acknowledgements

I owe thanks to many people, all of whom have helped me learn and attain some small amount of understanding in a field in which I received no formal training. I first and foremost thank those astronomers, historians, and other scholars who I have relied on to help me through the thickets of the material. I am a scholar who has wandered out of my own academic backyard and into theirs, and they have displayed admirable patience with my fumbling, misunderstandings, and ignorance. While there have been many people who have guided and assisted me in this project, I want to point out a few in particular. I would like to thank Cheri Adams and the staff of the Astronomy Department at the Boonshoft Museum of Discovery in Dayton, Ohio, for their wonderful public outreach and for helping to rekindle an interest in astronomy that had been dormant for years. Thanks also to the Miami Valley Astronomical Society, Mound Science and Energy Museum, and the Northern Colorado Astronomical Society, whose members offered me valuable feedback on various parts of what became this book. This book would have been impossible without the hard work and dedication of the people who preserve the amazing historical sites and artifacts of our shared human history, including the Cahokia Mounds Museum Society and the SunWatch Indian Village staff—especially site manager Andrew Sawyer and Curator of Anthropology Bill Kennedy, who saved me from a few boneheaded mistakes, the members of the Archaeological Survey of India tasked with maintaining such a difficult site in a difficult city as Delhi's Jantar Mantar, and the staff of the Adler Planetarium in Chicago, one of my favorite places on the Earth. Thanks also to my friend Leanne Martin, whose conversations about cultural astronomy through the years have helped me to think through much of this, and her husband Christopher Martin of Caltech, whose conversations about the GALEX discoveries and enthusiasm for astronomical discovery inspired me to conclude the section on contemporary astronomy with an account of his work. I would also like to thank a reviewer for Springer, whose insightful comments and suggestions both saved me from a number of (sometimes egregious) mistakes and helped me to further think through a number of issues.

Finally, thanks to my two sons, Siddhu and Francis, whose curiosity about the stars motivated me to learn more in order to teach them and encourage them not to lose that fascination. This book is dedicated to my parents, Gustavus Arius McLeod and Mary Alice McLeod, who instilled my love of learning, and who never gave up, even though they had every right to.

Contents

Chapter 1
Introduction

I am an eager student of the history of astronomy in general. I've devoured books on the greats of the West—Ptolemy and Copernicus, visionaries such as Brahe and Kepler, geniuses such as Newton and Herschel, and the revolutionaries of modern astronomy such as Einstein, Hoyle, and others. But through most of my years of study I had hardly ever encountered any information about astronomy done outside of the European and American West. In the histories of astronomy I encountered, Non-Western astronomy was generally relegated to a few paragraphs at best, about how "pre-scientific" cultures viewed and understood the sky. It would be easy to conclude from such reading that the scientific study of astronomy began in and was limited to the West. Strangely enough, it was my reading in *philosophy* that opened up to me the world of Non-Western astronomy and showed me its true extent and its purpose.

Years ago, I read a book by Zhou Guidian, Professor of Philosophy at Beijing Normal University, on Qin and Han Philosophy. One of my areas of specialization within philosophy is Han dynasty Chinese philosophy. On the basis of some work I have done in this area, I was been approached by a colleague in China to write a review for a professional journal on Professor Zhou's book. I took on the task, and during my reading, I came across the following line:

> In order to fully understand traditional Chinese philosophy, it is necessary to understand traditional Chinese astronomy.[1]

At the time, even though I had been an enthusiast of the history of philosophy as well as an amateur astronomer for a long time, I thought this was completely wrongheaded. We can, I thought, understand and engage with different modes of thought and different areas independently. Indeed, in my review of Professor Zhou's book I wrote something along these lines. I soon began to think more critically about this view, however. For some reason Zhou's quote about astronomy stayed with me well after I finished the review and had moved on to other issues and other works. I occasionally reflected on it, and thought about the divisions in our own academic world. We often see areas of thought as discrete and separate

[1]Zhou 2006, 190. Translation is my own.

A. McLeod, *Astronomy in the Ancient World*,
Historical & Cultural Astronomy, DOI 10.1007/978-3-319-23600-1_1

entities, around which there can be solid concrete walls, and no one outside shall enter. It is unheard of for philosophers to engage in astronomy or vice-versa. But was this the case for the ancients? Maybe we should not think of them as sharing our view of the autonomy of distinct areas of human thought, and holding that they should not "contaminate" one another. To what extent, I began to think, did the contemporary assumption of the intrinsic separability of different areas of thought affect my view of Zhou Guidian's claim about astronomy in classical Chinese thought? Because I am an academic and was trained in the same way as other contemporary academics, I wondered to what extent I had simply been acculturated to this view of the separability and distinctness of the areas of human thought.

I began to consider philosophical thought and astronomical thought, across a number of cultures. I asked myself the question: is it *true* that there is a link between philosophy and astronomy in world cultures, and is it the case that we can or ought to think of them separately? The idea of the separability of the sciences, and physics and astronomy in particular, may simply be part of the intellectual baggage of the modern world. There have been many excellent studies devoted to understanding the origins and rise of modern science. One of the most interesting and relevant lines of investigation that I have encountered concerns the cleft between science and the humanities advocated by figures in the 17th and 18th centuries. Ofer Gal and Raz Chen-Morris argue (Gal and Raz, 2012) that naturalization of the senses through creation of and analogizing to instrumentation in part created this contemporary epistemology of the sciences, which separates the "objective" and scientific from other aspects of human understanding. Rhonda Martens argues that Johannes Kepler's revolutionary advances in astronomy, still accepted today, were the result of his philosophical ideas which grounded and justified his astronomical work (Martens 2000). Akeel Bilgrami discusses the origins of a particular way of thinking about "naturalism" that rejected the normativity beginning with Isaac Newton and the Royal Society (Bilgrami 2010). All of this (and more) gives us good reason to think that the foundations of astronomy as we know it are by no means necessary, purely fundamental, or objectively justified.

Many contemporary scientists and others do not consider the ways in which our scientific understandings of the world are culture-bound, based on assumptions rooted in particular historical, philosophical, and cultural interests, and thus nowhere near as unquestionable and fundamentally necessary as they are often taken to be. Given that our conception of astronomy is fundamentally grounded by these assumptions connected with particular movements in Europe in the 17th and 18th centuries, it is natural that, when offering histories of astronomy or considering the uses of and understanding of astronomy in the past, scholars focus on that which can be understood as relying on the *same* or similar enough philosophical foundation. Thus, in most histories of astronomy, we see discussion centering on 17th and 18th century Europe, with some forays into older astronomical thought believed to have some connection in spirit with it, such as the astronomy of the ancient Greeks. In recent years, scholars have admirably attempted to add other voices outside the West to expand our understanding of the history of astronomy. But these additions of astronomical thought from China, India, Africa, and the

Americas to standard histories of astronomy have been problematic. The problem is that we have to understand that the background philosophical assumptions in these astronomical traditions are often very different than those of contemporary astronomy rooted in 17th century European thought. If we don't do this, then we will have to consider and appraise these traditions by the standards of the assumptions of contemporary astronomy, under which they can necessarily make little sense. These traditions then become reduced to toy systems of little significance. And this is often exactly how they appear in histories of philosophy. When we approach different traditions of astronomical thought we cannot evaluate and attempt to understand them using the same epistemological framework behind contemporary astronomical thought. That is, if we do not attempt to understand other traditions of astronomy around the globe in (as far as possible) terms of their own epistemological frameworks, we will not only fail to understand these traditions, we will also fail to see what is useful in them, and fail to *learn* from them.[2]

My own investigation of early Chinese astronomy and the astronomy of other cultures led me more and more to reject both the notion that astronomy can and should be separated from philosophy and other aspects of human understanding, and also that we can and should investigate and appraise ancient astronomical systems of astronomy as if they are contemporary systems of astronomy grounded in the same philosophical and cultural assumptions. All of this led me to seek out works on astronomy in ancient China and other ancient areas in the ancient world, including India, Africa, and the Americas, in order to try to better understand these astronomical systems and the philosophical foundations of them. It was not easy to find much information on the history of astronomy in these parts of the world. The first book I read that really opened my eyes to the extent and caliber of astronomy being done in these ancient Non-Western cultures was Volume 3 of Joseph Needham's *Science and Civilization in China*, in which he gave an account of the history of Chinese astronomy. Right away, I recognized the amazing sophistication of early Chinese astronomical systems, and the extent to which the Chinese had been far beyond anyone else in both their ability to observe as well as their understanding of the sky in the ancient world, and remained so for much of recorded history! I also learned that many of the observational innovations of people such as Tycho Brahe, which made modern Western astronomy possible, were likely derived from much older Chinese originals. It may have been the case that not only was Chinese astronomy enormously sophisticated and powerful, but also that it made modern Western astronomy possible!

I began to investigate astronomy in other ancient cultures, including those of Mesoamerica, after being introduced to the amazing and unique astronomy of the

[2]Gary Urton, in his discussion of Quechua mathematics, offers a good example of how, even in the realm of mathematics these cultural and philosophical assumptions are crucial to understanding a particular tradition: "numbers are not conceived of in Quechua ideology as abstractions whose nature and relations to each other rely on the predications of pure logic, as in the West. Rather, numbers are conceptualized in terms of social—especially family and kinship—roles and relations." (Urton 1997, 13).

Maya in a program at the Boonshoft Museum of Discovery in Dayton, Ohio. I then discovered the works of archaeoastronomer Anthony Aveni, whose *Skywatchers of Ancient Mexico* opened my eyes to not only the amazing and advanced astronomy of the ancient Maya, but also to the field of archaeoastronomy in general. Once I discovered how amazing and important these astronomical systems of thought were, I jumped into reading about a number of systems, and discovered a treasure trove of material on them, including that of the Mississippian culture of the city of Cahokia.[3]

The more I investigated the astronomical thought of numerous world cultures, the more convinced I became that I had been wrong to criticize Zhou Guidian's claim, that ancient Chinese astronomy and philosophy were intimately connected. I also came to see that I had been woefully uninformed about the history of astronomy, having seen it as primarily a Western construction, its story playing out solely in Europe and later in European America. Indeed, I came to believe that I had been wrong about a fundamental assumption I had made concerning the academic endeavor in general, one that I had been indirectly taught during my training in philosophy—the idea that we should or even can keep separate disciplines such as philosophy, history, and astronomy. I have come to see each of these areas as far less distinct and unique than I used to. Doing history and doing philosophy can overlap, and can indeed sometimes be indistinguishable. The same goes for astronomy. Indeed, when I began to investigate the astronomy of the ancient cultures of the Maya, Mississippians, and Chinese, I saw emerge a pattern of thinking about nature itself that challenges this view of separation and boundary inherent in much of the thought of the modern Western traditions. Zhou Guidian's claim is, I discovered, no less true for the ancient Maya, Aztecs, and the people of Cahokia than it is for the ancient Chinese. I began to wonder what a synthesis and attempt to understand this philosophical principle in connection with the astronomical thought of these peoples would look like.

Today, when I walk into a museum of science and astronomy, I hardly ever see mentioned (other than occasionally as an aside) the rich history of non-Euro-American astronomy, even though the story of the human engagement with astronomy is a story most of which takes place outside of Europe. One might, in an American museum, see brief mention of Native American sky watching or constellation myths. Outside of this, little is said. Wandering around our museums or reading our books on astronomy would leave one to think that the native people of this continent, as well as people almost everywhere else in the world except

---

[3]While there have been an increasing number of works written in this area, the available work still pales in comparison to that available on issues in the history of Western astronomy. Material on astronomy and cosmology in specific at Cahokia is limited, but a number of authors writing on Cahokia in general cover it. See in particular Chappell 2002, Fowler 1996, Pauketat 2009.

Europe, either had little interest in or little expertise in astronomy.[4] Neither of these is true. We've simply been held hostage for too long by our own unwillingness to look outside of the confines of Euro-America for the story of the history of astronomy. Not only can looking outside improve our understanding of other cultures and of astronomy, but it can also help us better understand the history of astronomy in the West, and its own background assumptions. It is impossible to recognize in our own thought what is cultural or philosophical assumption and what is universal if we don't have a good understanding of other traditions of thought. The tendency then becomes to accept *all* of the foundations of our own astronomical thought as necessary and universal.

This book is in part an attempt to encourage deeper study and greater understanding of ancient and Non-Western astronomical traditions. I focus on a few important skywatching cultures, all of which developed sophisticated observational techniques, predictive abilities, and cosmologies. I compare the astronomical thought of these traditions to contemporary understandings of the phenonomena they were concerned with, and finally with Western astronomy as a whole. The book ends with a consideration of Western astronomy, in the hope that once we have seen some of the assumptions and concerns of ancient and Non-Western astronomical systems, we will be able to more clearly see the contingent nature of astronomy in the West, and recognize some of the philosophical assumptions underlying it. The main focus is on four particular areas of the world, East Asia (current day China), South Asia (current day India, Nepal, Pakistan, and Bangladesh), the North American plains (current day USA, surrounding the Mississippi River), and Mesoamerica. These stories need to be added to our shared human narrative concerning the history of astronomy. In museums charting this history, we ought to see plates on the Mayan calculations of Venusian orbits and phases alongside the Western developments in understanding our neighboring planet. We should see discussions of the ingenious and unprecedented Chinese observational system alongside discussion of the revolutionary observational program of Tycho Brahe, many of whose innovations had already been in use by the Chinese for thousands of years. We should see Indian mathematics and cosmology placed beside the early Ptolemaic and Copernican cosmologies of the West, beside which the Indian systems would fare quite well.

Scholars should also, I believe, devote much more time and resources to investigating the history of influence between Western and Non-Western systems of astronomy. What has traditionally been told as a Western story, with its origins in Greece and traveling into Western Europe, turns out to be a much broader and more diverse story. What would stop ways of thinking from crossing the massive border between Western Europe and the Slavic world of Eastern Europe, for example?

---

[4]Although there is still not as much Non-Western material as desirable, some museums do relatively very well at this, and attitudes are slowly beginning to change about the importance of Non-Western philosophy. One of my favorite museum collections in astronomy, for example, of the Adler Planetarium in Chicago, IL, contains a large number of Non-Western astronomical tools, and magnificent exhibits on these instruments.

That between Eastern Europe and the Central Asian Steppe? That between the steppe and the rest of East and South Asia? Especially when we are willing to accept that the developments and advancement in mathematics, philosophy, and science of the Arabic-speaking world passed without much of a problem into Western Europe, over a certainly harder-to-penetrate divide of water, mountains, religion, and ethnicity (possibly because there is no choice given how explicitly the medieval texts admit their reliance on Arabic materials), why are we so unwilling to accept that there may have been mutual transmission between the West and more geographically and culturally accessible areas?

Why are we unwilling to consider the high likelihood of mutual exchange in the Americas as well? The ideas of the later Renaissance (14th–17th century) corresponded to the time of the European entrance to the Americas (15th–16th centuries), and the explosion of new concepts and ways of thinking in Europe cannot be, no matter how strained of wishful thinking one engages in, simply a result of sudden and inexplicable innovations of the Europeans alone. Why wouldn't such a thing have manifested itself long before the Renaissance? The ideas the Europeans adopted and adapted from the rest of the world were not, and to this day still mostly have not been, properly attributed. Much of the Renaissance was a compilation of the ideas of others without citation (to use academic imagery), leaving future generations to believe (as perhaps was intended) that the European makers of the Renaissance came up with all of this on their own.

This is not to say, of course, that people in Europe did nothing of their own. Certainly, there were brilliant astronomical, philosophical and other developments in Europe, just as there were in other parts of the world. But the idea that Europe was the sole origin of all important developments in astronomy, science, and philosophy is simply false. In our 21st century world, we ought to have enough distance and enough perspective to see through this untruth. The historical and archaeological material and data is there—we simply have to investigate it. Indeed, even many scientists and philosophers in Europe *admitted* their debt to "Non-Western" thought. The thinkers of the medieval period of Europe were most willing to admit their dependence on the work of other non-European people, including Arabic sources created by a "heathen" people who were supposed to be their mortal enemies. It is likely that this is the only reason the traditional Renaissance story of philosophy, for example, admits the contribution of Islamic philosophers in the middle ages. The Medieval European philosophers themselves recognized and explicitly discussed their debt to the Arabic philosophers. But suspiciously, after the discovery of Plato and Aristotle and the intellectual digestion the works of the Arabic philosophers, the Middle East strangely goes silent again in the traditional Western story, as if all of a sudden late in the first millennium the Arabic speaking world produced philosophical thought, then as soon as they had done their job of transmitting this to its rightful European heir, they ceased doing philosophy, vanishing back into the abyss from whence they came.

This story is of course absurd, and no one should be duped by it. How could something like this happen? How could a culture who had never heard of or seen or thought about philosophy before suddenly begin doing philosophy at the highest level, then a few years later simply forget everything and fall back into their pre-intellectual state? Of course this could not, and did not, happen. Philosophy was done in the Middle East long before and long after Avicenna, Averroes, and Al Kindi. And so was astronomy. Both in the Middle East, and elsewhere in the world

There are still of course many unanswered questions about the relationship between Western and Non-Western astronomical thought. Why, for example, did the Chinese, for all their observational genius, never produce a thinker like Kepler who synthesized all of this excellent data to make sense of and offer a simple and mathematically powerful theory of the movements of the heavens? Perhaps a large part of the reason that the Chinese never produced their own Kepler was the same reason that they were such excellent observers. To produce a thinker like Johannes Kepler, a strange confluence of things was necessary: an intellectual culture that held to the view that the cosmos had to obey some kind of order (the intellectual heritage Europe gained from Aristotle's "perfection of the cosmos" ideas), combined with a realistic and detailed observational ability that took seriously and did not avoid the difficult problem that the cosmos does *not* appear orderly. It is the latter that simply did not exist in the West until the time of Tycho Brahe. And thus, perhaps the greater miracle was not Kepler, but Tycho Brahe. Europe had no shortage of bright young men looking to make sense of the fundamental order of the cosmos. But one willing to challenge the notions of the fundamental perfection of the heavens on the basis of observation, as did Tycho with the comet of 1577, and the supernova of 1572—such a person was truly rare in Europe. It was as if a Chinese astronomer had entered Europe at a critical juncture, when Copernican thought was beginning to hold sway. Only at the confluence of the two disparate systems of thought could such a revolution in astronomy happen. As strange as it seems, China and Europe in some way met inBrahe and Kepler.

Anyone who knows the history of astronomy has heard the stories of the major European figures: Copernicus, Galileo, Brahe, Kepler, Newton, etc. The story of astronomy and its impact on our world is generally told as a European story, through the eyes of European men. But they were neither the only ones to cast their eyes and minds on the skies, nor did astronomy play the role in their culture that it did in a number of others. Other parts of the world, particularly the Americas, China, and India, as we will see, had a much longer and more significant early history of astronomy than did Europe. Thus, the story I begin to tell in the following pages is the greater part of the overall history of astronomy (just as those outside of Europe and America make up the vast majority of the population of humanity). It is a story of our shared human cultural heritage, a story of scientific innovation, as well as mythology, cosmology, and religion. And it is not a story of *them*—it is a story of *us*.

In each of the four parts of the book, I recount some of the history and main features of an early astronomical system of thought, usually surrounding some major event in the history of astronomy, and then I move into a discussion of our contemporary understanding of the celestial phenomenon in question. I hope to show that in our considerations about the heavens today, we are not intrinsically different from these cultures far removed from us in time (and sometimes) in place.

The ways we arrive at our own understandings of the sky are very similar to the ways the ancients arrived at theirs. And the astronomical thought of the Maya, the ancient Chinese, or the Vedic thinkers is just as relevant to us today as is the thought of Galileo, Copernicus, or Kepler. These are our intellectual ancestors, anticipating and often even eclipsing later developments recognized today as watershed moments in the history of astronomy. It is time that we give these forgotten astronomers their due in our histories, making them more than just a footnote or a mention in a preface, on our way to discuss more deeply the Western history of astronomy. It is time to make these amazing astronomers known again.

References

Bilgrami, Akeel, 2010, "The Wider Significance of Naturalism: A Geneological Essay", De Caro, Mario and David Macarthur (eds), *Naturalism and Normativity* (New York: Columbia University Press)

Chappell, Sally, 2002, *Cahokia: Mirror of the Cosmos* (Chicago: University of Chicago Press)

Fowler (ed), 1996, *The Ancient Sky and Skywatchers of Cahokia: Woodhenges, Eclipses, and Cahokian Cosmology*, *Wisconsin Archaeologist,* 77 (3-4)

Gal, Ofer, and Chen-Morris, Raz, 2012, *Baroque Science* (Chicago: University of Chicago Press)

Martens, Rhonda, 2000, *Kepler's Philosophy and the New Astronomy* (Princeton: Princeton University Press)

Pauketat, Timothy, 2009, *Cahokia: Ancient America's Great City on the Mississippi* (New York: Viking-Penguin)

Urton, Gary, 1997, *The Social Life of Numbers: A Quechua Ontology of Numbers and Philosophy of Arithmetic* (Austin: University of Texas Press)

Zhou, Guidian, 2006, *Qinhan Zhexue* (Qin and Han Philosophy), (Wuhan: Wuhan Chubanshe)

Part I
Watchers of the Sun

Chapter 2
The Americas

Cahokia

The mounds scale out over the horizon almost as far as one can see across the plain. Climbing to the top of the Monk's Mound, the largest mound of the complex, one rises well above the trees, and can see for many miles around. You can see the famous Gateway Arch of downtown St. Louis in the distance, and the Mississippi Valley for many miles in both directions. And most clearly and distinctly, you can see the *sky*. The entire dome of the sky becomes visible atop the enormous mound, with the all-important horizon discernible and unobscured. One of my first thoughts when I scaled the stairs leading to the top years ago and glanced out at the view was "this place would be a great spot for astronomical observation!" The scope of this present complex is amazing, even in its dilapidated state after so many years since its precipitous abandonment in the mid 14th century CE. At its height the city must have been a thriving metropolis. But the question continually arises–what was the purpose of these grand areas, the mounds, the astronomically aligned structures around the entire city? The people of Pre-European America understood and had a great concern for celestial phenomena. The movements of the sun and the planets were of enormous import to them, as they were for civilizations elsewhere in the world. But what was their specific concern with the sky? How did they understand its significance, and what role did it play in their thinking about themselves, their place in the world, and their destiny? Clearly, the people who built and maintained Cahokia had a greater concern for and care about astronomy than do those in contemporary urban society, for example (Fig. 2.1).

When Euro-Americans first saw Cahokia in what is today the Illinois plains of the Mississippi Valley, it had been abandoned for years. This massive complex was a ghost town. It didn't take long to uncover the amazing extent of this ancient city. It had been a metropolis–and it was now empty. Henry Marie Brackenridge, a lawyer, author, and later judge and US Congressman from Pennsylvania, was the

A. McLeod, *Astronomy in the Ancient World*,
Historical & Cultural Astronomy, DOI 10.1007/978-3-319-23600-1_2

Fig. 2.1 Monk's Mound, the largest of the structures at Cahokia, still stands tall even after hundreds of years of neglect and wear. The mound likely began as a burial mound in its earliest days, and its ceremonial significance led to its "repurposing" as a focal point for ceremony and the ruler. The residence of the ruler of Cahokia once stood atop the mound, and many major ceremonies of the city were performed from there. The name 'Monk's Mound' is completely unconnected to the Mississippians, rather it derives from the Trappist monks who inhabited the site of Cahokia for a few years in the early 19th century. (Photo credit: Cahokia Mounds State Historic Site)

first known to write about the site, in a letter to former president Thomas Jefferson in 1813. Brackenridge writes in effusive prose about the site he visited in 1811:

> Nearly opposite St. Louis there are the traces of two such cities, in the distance of five miles, on the bank of the cohokia [sic], which crosses the American bottom at this place. There are not less than one hundred mounds, in two different groups; one of the mounds falls little short of the Egyptian pyramid Myrcerius. When I examined it in 1811, I was astonished that this stupendious [sic] monument of Antiquity Should have been unnoticed by any traveller… (Looney 2009)

It had been as if the ancient American equivalent of New York City emptied out and was abandoned, to be regained by the earth and the overgrowth of years. When these explorers and archaeologists first saw the gigantic mounds dotted throughout the city, they assumed they must be either burial mounds or have some kind of religious or political significance–perhaps the living areas of the rulers of the city? Cahokia was, and in many ways remains, a mystery.

Cahokia was not the only megalopolis to emerge from the reclamation of the earth in the Americas. Nor was it the only one with a key concern with and link to astronomy, the observation of the cosmos. But Cahokia has a particular interest in its link with the heavens, as it was an unusual and itself strange celestial event that led to a major change in its society and life.

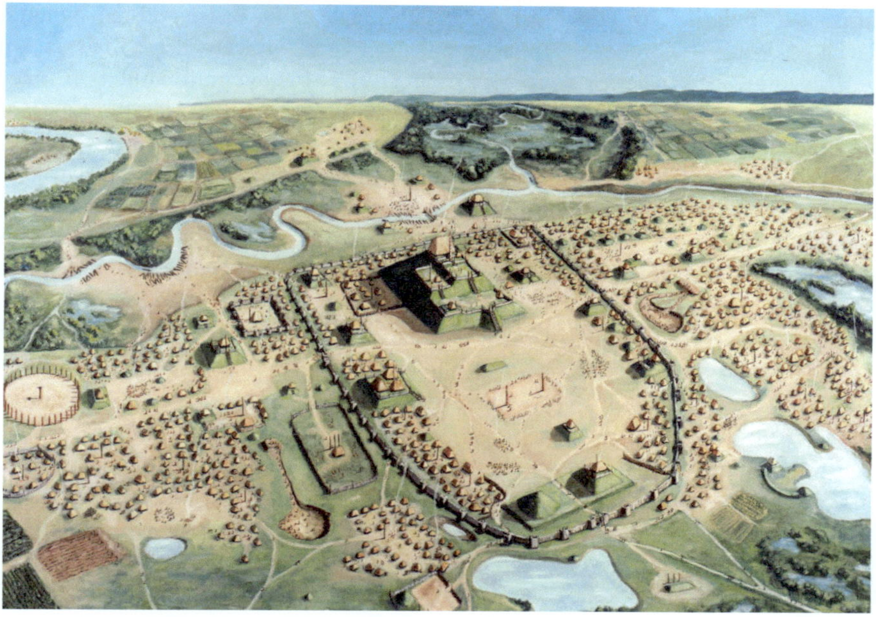

Fig. 2.2 Artist's rendering of what "downtown" Cahokia may have looked like at the city's peak, with Monk's mound in the upper right. (Photo credit: Cahokia Mounds State Historic Site)

Archaeologists, having worked on the Cahokia site for years, determined that a major restructuring of society in the Mississippian town took place in the mid 11th century CE.[1] The "old city" of Cahokia that existed on this site prior to this period had been more a village than a city—a small hamlet of perhaps a couple thousand people at most. Something happened in the mid 11th century that caused people in the region to radically transform their city. They built enormous mound pyramids and other earthen structures in a very short time, and the population exploded almost overnight. Cahokia went from sleepy Mississippi Valley village to massive urban center. And the planners had intended just this—the construction of the new city of Cahokia was a planned urban project unlike anything that had been seen anywhere in North America. The growth of Cahokia was not the slow and natural growth of a city that comes with steady migration. Rather, it was a centrally planned decision. The builders of the new city of Cahokia had every intention of constructing a new metropolis (Fig. 2.2).

Of course, there are many pressing questions. Where did the builders of Cahokia come up with or learn the idea of massive urban ceremonial centers? How did they

[1]There is still disagreement about just what kind of restructuring took place and how it did, though there is more agreement about the time period within which it took place. Timothy Pauketat's view lines up most closely with the hypothesis I consider here. He discusses this "Big Bang" view of Cahokia's rise in Pauketat (1997, 2009).

attract a population so quickly? How did the idea of the large earthen pyramid develop? Was it an independent regional discovery, or were the Mississippian peoples influenced (whether directly or indirectly) by the pyramid-building peoples of Mesoamerica? Perhaps an even greater mystery surrounds the startling date of the construction of the new city of Cahokia: The years surrounding 1050 CE.[2] Archaeologists determined through carbon dating of the features of the new city that it was around this date that the massive transformation of Cahokian society took place. This date may seem meaningless to most of us, but some familiar with astronomy will recognize the nearness of this date to a very famous astronomical event: the great Supernova of 1054 CE.

Only one account of the massive celestial explosion that took place in 1054 comes down to the modern day–that of the meticulous observers in the court of the Chinese Song dynasty.[3] Strangely, there is no written account anywhere else in the world of the massive 1054 supernova. We know today that the event must have been obvious to those in the northern hemisphere with clear enough skies—the 1054 supernova would have been the brightest object in the sky behind the Sun and Moon–but there are no European or Middle Eastern accounts of the event. The "new star" may indeed have caused a stir in the Americas, though. While much of the discussion of the significance of the 1054 supernova in the Americas is speculative, and it will likely be impossible to ever know just how significant the event was for the various peoples of the continent, there are some suggestions that it was seen as important. A number of still existing etchings and paintings in the American Southwest are believed by some to depict the unique event, including a painting at Chaco Canyon in modern day New Mexico of the state of the night sky in the year of the supernova's appearance,[4] and a number of pottery pieces from the Southwest (Mimbres Valley, New Mexico) that date to around 1050 seem to illustrate the supernova. Other possible depictions elsewhere in North America, including a possible alignment to the object in the famous Serpent Mound in southern Ohio, attest to the fact that not only did the people of this region take likely notice the startling celestial event, but that they may have taken it to have an unusual significance (Fig. 2.3).

And why wouldn't they? Naked eye visible supernovae, the brightest and/or closest of exploding stars, often so bright they are visible during the day when they

[2]About A.D. 1050, the American Bottom experienced the political and economic equivalent of the Big Bang… The event brought about the abrupt and large-scale transformation of community order, the physical landscape of Cahokia, and the entire northern expanse of the American Bottom floodplain. (Pauketat 1997, 31–32).

[3]The appearance of the supernova is recounted in two sources, the *Song shi* ("History of Song") and the *Song huiyao jigao* ("Compendium of Documents of Song").

[4]Scholars disagree over whether these inscriptions depict the 1054 supernova, rather than some more ordinary object such as the planet Venus.

Fig. 2.3 Detail of the petroglyph from Chaco Canyon (New Mexico), representing what some suggest may be the supernova of 1054 CE, beside a crescent moon and the "Hand" constellation. The hand constellation, a common grouping of stars among the indigenous peoples of North America, consists of what many westerners know as the "Orion" constellation, including Orion's belt, the feet, and the Orion Nebula

explode in our galaxy, are among the rarest events in the skies.[5] There have only been a handful of such events in all of recorded human history. No one alive today has ever seen such an event, nor have even any of our great-great grandparents. The last time such an event happened was 1602. Some lucky generations, however, were showered with an embarrassment of riches. Many people who witnessed the great supernova of 1572, for example, would have still been alive when the supernova of 1602 appeared in the sky. Johannes Kepler, often considered the father of modern astronomy, lived through both of these events, even though he was a mere infant in 1572. Kepler himself thus didn't have much to say about the 1572 supernova. This supernova is often associated with the man who would become his mentor and predecessor, Tycho Brahe, whose measurements of the supernova definitively demonstrated to the western world (as we will see later) that Aristotle's theory of the perfection and (thus) unchangeability of the heavens was false. Kepler himself played a large role in the analysis of the 1602 supernova, which is today often referred to as 'Kepler's Supernova'. Kepler's observations of this "new star" were recorded in his book devoted to it, *De stella nova in pede Serpentarii*, of 1606.

When supernovae happen in our galaxy, they are hard to miss for anyone with even the most rudimentary familiarity with the night sky. And they are so rare and magnificent that people tend to remember them, memorializing them in things like pottery and cave paintings, astronomical books or detailed court records (as in the case of the Chinese astronomers, who we will meet in the next section).

[5]Although it is still somewhat curious that the 1054 supernova would create such a stir, given that there had been an even brighter supernova, indeed the brightest in human history, that had appeared in the skies of the northern hemisphere a mere 48 years before, in 1006 CE.

The people who inhabited Cahokia are known today as the "Mississippian" people—a name that would not have been used by these people themselves, but that was given to them much later by scholars because of their range of cultural diffusion in the wider Mississippi Valley region. Cahokia was the core of this civilization, a megalopolis that would have dominated the wider culture and politics of the region and the people throughout this part of the continent. The mystery of Cahokia is almost irresistible to those with an interest in pre-Columbian America, or just anyone with a curiosity about seemingly inexplicable phenomenon. Not only this, but Cahokia, its meteoric rise, and its puzzling abandonment may, if we learn the facts about its ultimate fate, be able to teach us important lessons about urban population, growth, rise and decline in our modern world. These are lessons that no doubt will be of utmost importance, given the 7 billion plus population of our increasingly urban world.

I like to believe, although I certainly don't have irrefutable evidence for this, that the story of Cahokia's rise and that of Mississippian culture in its larger scope in general, is a story at least in part about astronomy.[6] And astronomy, of course, often has its basis in the religious, philosophical, and cultural life of a people. It is certainly not unheard of in pre-Columbian American civilizations for astronomy to hold a central significance. Ultimately, regardless of the cultural impact of SN 1054 (if any), in various cultures in the "New World" we see well-documented concern with celestial phenomena, well before any visitors from Europe arrived.

Mesoamerica

Astronomy in the Americas has a rich history, as the detailed observation and study of celestial phenomena took place in almost every inhabited area of the two continents. The most well-known and perhaps best developed (or at least best recorded for posterity) astronomy of the ancient Americas was that of the Mesoamerican empires, particularly the Maya and the Nahua (Aztec). One can still see well-preserved astronomical artifacts from both cultures throughout Central America.

The Maya are perhaps best known to many of us for the so-called doomsday prediction of the end of the world on December 21, 2012—the apocalypse that wasn't, as I'm sure you remember. Unfortunately for the doomsayers, but fortunately for the posterity and good name of the Maya astronomers, this prediction was not only misinterpreted, but was actually never even made by the Maya at all. No Maya astronomical calendar makes any such prediction of the end of the world, on

[6]Timothy Pauketat, an archaeologist at the University of Illinois, and some other scholars, agree. Pauketat writes "The latest radiocarbon dating places the construction of New Cahokia at about 1050. The closeness of that date to the appearance of the supernova in 1054 has prompted some archaeologists and historians to question whether the astronomical event could have caused or somehow contributed to the momentous changes that took place in the Mississippi River valley." (Pauketat 2009, 23).

any date.[7] In addition, later popularizers of the "end of the world" claim both misunderstood Maya astronomy, and also basically invented the idea that the end of a *baktun* cycle in the Maya calendar corresponded to a claim about the end of time. On the contrary, the Maya view of cycles of time is one of *repetition*, following the cycles and sequences we witness throughout nature. Indeed, it would not be a *cycle* at all if it did not repeat. The Maya no more held that the world would end at the finish of *baktun* cycle previous to the present one (not December 21, 2012, as some misinterpreted, but close) than a contemporary would hold that the world will end at the end of this calendar year (the end of our 12 month cycle which then begins anew) (Fig. 2.4).

The beginning of the last *baktun* cycle in the Maya calendar, and its related ancestors following the "Long Count", was August 13, 3114 BCE, a date with almost mystical significance in Mesoamerican culture, due to the profound events associated with this date.[8] There is some similarity here to our own Common Era (CE) count–or *anno domini* (AD), depending on one's ideology or preferred calendric terminology, generally held to correspond with the birth of Jesus. There have been arguments that the first civilization to adopt the Long Count calendar was the Olmec civilization centering around what is today La Venta in the state of Tabasco on the Gulf Coast of southern Mexico. According to this view, Olmec culture formed the basis of later Mesoamerican imperial culture, influencing the cultures that later became the Maya and last the Aztec civilizations, both of which adopted similar calendric systems.[9] Others contend that the calendar had its beginnings not on the Gulf Coast, but further south, on the Pacific Coast in the region of current day Oaxaca.

The mysterious city of Teotihuacan in modern day central Mexico (about 30 miles outside of Mexico City) is constructed such that it's main avenue, the Avenue of the Dead (*avenida de los muertos*, a Spanish translation of the Nahuatl *Miccoatli*) was aligned to correspond with sunset on the day of August 13.[10] I call this city mysterious because it is unknown exactly which peoples were associated with the site in its earliest stages, even though it later fell within Aztec influence–and indeed, the name 'Teotihuacan' itself is an Aztec name, given to the city long after its original inhabitants had abandoned it (for still unknown reasons). There are varying

---

[7](Schiele and Friedel 1990) make short work of this fictional end-of-time claim concerning 2012, as do a number of other Mayanists.

[8]This is based on the proleptic Gregorian calendar and the Thompson calendar correlation. According to the Goodman-Martinez-Thompson correlation, the date is August 11, 3114 BCE in the Gregorian calendar. In the Julian calendar, the dates are September 8, 3114 BCE (Thompson) and September 6, 3114 BCE (GMT), Julian Day 584285, where Julian Day is defined as days elapsed since January 1, 4713 BCE (Julian).

[9]Diehl 2004. While the Aztecs *did* not use the Long Count calendar, they did use an equivalent to the *tzolk'in*, called the *tonalpohualli*, which along with the 365 day *xiuhpohualli* calendar (roughly equivalent to the Maya *haab*), formed the "calendar round" of 52 years.

[10]Vincent Malmstrom convincingly argued for this position against previous views according different alignments to the Pyramid of the Sun (Malmstrom 1978).

Fig. 2.4 Depiction of a initial Series from Quirigua Stela C. The topmost glyph is referred to as the Initial Series Introduction Glyph (ISIG), and the glyphs below record the date in the Long Count as well as the *tzolkin* and *haab* calendars. The date shown here is 13.0.0.0.0, 4 Ahau 8 Cumku, corresponding to the Gregorian calendar date of August 11, 3114 BCE—the purported (mythical) origin date of the calendar and beginning of the 13th baktun, which ended in December 2012

views about the identity of the original inhabitants of the city, as well as its cultural influence and forebears. Still, the August 13th alignment seems compelling (Fig. 2.5).

Clearly, this date was important to the civilization at Teotihuacan as it was for the later Mayans and Aztecs. As to why this date was selected as the beginning of the current era of the Long Count, we can only speculate. The view of Olmec origins of the calendar is controversial, and any purported importance of August 13th in that culture would be even more speculative. However, presumably many different Mesoamerican peoples saw the date as a highly significant one, given that it dictated not only the beginning of their calendar, but also the design of their cities.

Fig. 2.5 A model reconstruction of the "Pyramid of the Sun" at the ancient city of Teotihuacan (just outside of current day Mexico City). The pyramid was a major ceremonial center for the people of Teotihuacan (whose identity is still debated). The pyramid, built around 100 BCE, appears to be aligned along its main stairway and altar with the point of sunset on August 13, the origin date of the last series of the Long Count calendar. Credit: Wolfgang Sauber

August 13th would come to have a less auspicious significance in much later years, as it signaled the end of the dominance of Aztec culture in Mesoamerica, when the Spanish explorer, man at arms, and *conquistador* Hernan Cortes and his troops captured the Aztec capital of Tenochtitlan (at the site of current day Mexico City) on August 13th, 1521, and took captive the Aztec emperor Cuauhtemoc, who was later executed. This signaled not only the end of the Aztec empire, but also the end of native rule in Mesoamerica (except for pockets of indigenous rule and resistance in the Maya region for years afterward), and the beginning of European colonialism.

When we consider the question of the astronomical alignment of major structures and city plans, of course, we must be careful, because in our fervor to discover astronomically significant alignments we may determine that completely accidental arrangements, or arrangements made with some non-astronomical purpose in mind, were purposefully astronomically conceived. There are only 360° in a circle, and on the horizon, after all, and thus even with completely random choice, there is about a 1 in 360 chance that any structure will be pointed toward any given significant celestial event along the horizon. This ratio increases significantly when we take into account all of the different significant celestial events that might be recorded along the horizon. And certainly even a people very concerned with astronomy in general might have different principles guiding them in building their cities and structures.[11] There are a number of reasons that we can take the peoples of Mesoamerica in general to have had a deep astronomical concern, one that translated into the way they laid out their cities and organized their lives. This concern was shared in North and South America as well. The philosophy of nature underlying and grounding the concern with the heavens that we see in Mesoamerica is one we tend not to share today, and one that allows us to draw a heavy line between the sky and other aspects of our lives, between humanity and nature, and between disciplines and kinds of knowledge, for example. The cultures we investigate in this book thought, for the most part, very differently about nature, and the skies.

The Mesoamerican people most closely associated with the Long Count calendar by the rest of the world, of course, are the Maya, whose astronomical and general intellectual culture was perhaps the most highly developed of all the early Mesoamerican peoples. The Maya civilization, even though in some form it continues to exist today,[12] flourished mainly between 100 and 1100 CE, and was at its creative and technological peak during the so called "Classic Period", in the years from 250–900 CE. It spread from what is now southern Mexico, including the Yucatan peninsula area (also famous because it was just offshore from this peninsula, millions of years before that the massive asteroid that caused the extinction of the dinosaurs landed), south through Central America. The center of power of the Maya civilization was in the area of what is today the Mexican Yucatan and northern Guatemala, where we still today find the remains of such once magnificent cities as Tikal, Chichen Itza, Uaxactun, and El Mirador.[13] Maya culture reached a remarkable sophistication, equal to any of its contemporaries in the "old world", including sophisticated understanding of astronomy, and the creation of astronomical, political, religious, and philosophical texts. Sometimes

[11]As an example of accidental astronomical alignment, we can take the example of the layout of the city of Chicago, IL. Downtown Chicago's streets are laid out along east to west lines, and the buildings were constructed along the blocks following this pattern. Because of this, along with the fact that on the equinoxes the sun sets exactly due west, it turns out that on the equinox days, one standing along an east-west Chicago avenue will see the sun emerge from behind buildings and align perfectly with the avenue at sunset. Having fun with this coincidence, some have called it "Chicagohenge". But certainly the designers of the city of Chicago did not purposefully align their streets to correspond to sunset on the equinoxes—they were just working on a grid based on the cardinal points, and it just happens to be the case that the equinoxes are also correlated with the cardinal points.

[12]The Maya still live in the land of their ancestors, though they have largely adopted Spanish culture and have been subsumed into the nation-states formed in the breakup of the Spanish colonial empire of New Spain, such as Mexico and Guatemala. There are also many others throughout the region and elsewhere in the world with Maya ancestry. The height of Maya civilization may be past, but the Maya survive.

[13]The original names of these cities were different from those we know today. This is obvious in the case of El Mirador, but the Mayan names given to other sites do not match their ancient designations either. The city we know as 'Tikal' was most likely called 'Mutul' in the Classic Period, for example, and the city of Copan, in present day Honduras, was likely called 'Xukpi'.

people are surprised to hear the fact that the Maya possessed both writing and texts. Part of the reason for this is likely because most of the examples of Mayan language and writing we have access to today come from engravings on buildings or other stone objects.

Unfortunately, there are very few Mayan[14] texts remaining, including the glyphic texts etched on stelae and buildings, or painted on pottery and in a number of books, which are still not well understood due to the lack of additional material. But it did not have to be this way. The Maya once had a robust literary culture, much of which was destroyed by Spanish invaders in the name of religion and colonialism. Maya priests and elites (most likely) compiled a number of books on various topics, which were generally kept, as such texts tend to be in just about every culture, in places of political and religious significance in great cities. During the early years of Spanish colonization, Christian missionaries purged most of this material, in the belief that they were somehow saving the backward and blasphemous peoples of the Americas by consigning thousands of years of their learning, knowledge, and culture to the flames. The most egregious example of this was carried out by the Franciscan monk Diego de Landa, who had been sent to the Americas in 1549, after the Spanish conquered the Yucatan, to convert the natives to Christianity, and was made bishop of the new Archdiocese of Yucatan. Landa was upset and frustrated by the fact that many of his Maya converts continued to practice the religious rituals of their ancestors–which they apparently did not see as inconsistent with adopting Christianity. The situation was similar to that of the native north American peoples as well, as it had been with Europeans when they adopted Christianity about a thousand years earlier. Converted Europeans in the early years of Roman Christianity continued to celebrate festivals such as the *Dies Natalis Solis Invicti* (Birthday of the Unconquered Sun[15]) on December 25th, which church leaders decided to absorb rather than to resist, because people had a way of ignoring church injunctions to give up habits, festivals, and cultural practices they'd always known in order to please God. Thus, the Christian holiday of Christmas is celebrated to this day on December 25th. Similar situations can be seen in the earliest Christian transmission to gentiles in the relaxing of strict Jewish laws concerning circumcision, strictures against eating certain kinds of food, and even injunctions concerning sexual mores and divorce.

Regardless, Bishop de Landa saw the Maya refusal to abandon their historical practices as highly offensive to Christianity and in need of rectification. Part of his response to this was an *auto-da-fe* at Mani in 1562 in which a massive number of ancient Mayan texts and artistic images were burned, thus destroying a large part of

---

[14] I follow the convention here of referring to the people as 'Maya', and the language and texts of these people as 'Mayan'. Generally the latter refers to languages, such as the "Mayan language family".

[15] A Roman god associated with the return of the sun to the highest skies, and thus appropriately celebrated close to the winter solstice. The emperor Constantine, famous convert to Christianity, identified himself, and Christ, with Sol Invictus, and the imagery he adopted throughout his reign suggests this triune association.

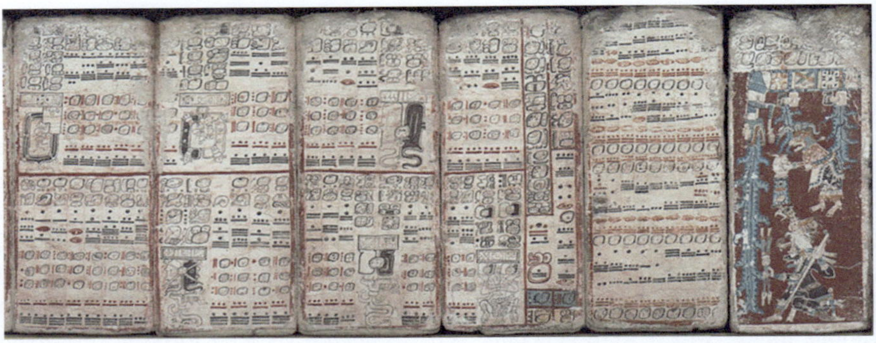

Fig. 2.6 The Dresden Codex, one of only four extant codices in the glyphic language of the pre-Columbian Maya, contains the most information of any of the codices on astronomy, including a venus table and lunar series

the cultural heritage of an entire people. Few texts survived the persecutions of de Landa and other likeminded Spanish missionaries, and today there are only four known texts to survive,[16] known to us today by the cities in which they are kept: the *Dresden Codex*, the *Madrid Codex*, the *Paris Codex*, and the *Grolier Codex*. For our purposes here, the Dresden Codex is the most interesting, as it contains astronomical material and also gives us some insight into the Maya view of the heavens and their significance.

While Maya glyphic writing and the early Mayan language of the Classic Period have still not been completely deciphered, much of it has been. There is a wealth of information about the Maya just within the extant codices, and we are only left to wonder what a magnificent amount we could have known about Maya culture (including astronomy) had the early colonialists been interested in preservation rather than religious conversion. Especially unique in the Dresden Codex is the role of the planet Venus in Maya astronomy. Venus seems to have had unique significance in a number of Mesoamerican cultures. There has even been disagreement as to whether the cave paintings, inscriptions, and other representations throughout the Americas seeming to depict the 1054 supernova may instead represent Venus (Fig. 2.6).

As mentioned above, the most famous astronomical association many make with the Maya is the Long Count calendar, which actually pre-dates the Classic Period Maya. The Long Count was likely created by a Mixe-Zoque people such as the Olmec, as the earliest known use of the calendar was in this area. The exact line of transmission is not known. The Long Count may have been known at Teotihuacan as well, given the site's alignment to the formative date of August 13.[17] Although the calendar was not created by the Maya, the Maya did *perfect* the calendar that

[16]Although there are other partial, damaged, or minor works discovered.

[17]The significance of August 13 as a formative or otherwise important date of course may be wider in the larger Mesoamerican region than the Long Count calendar.

they adopted from other peoples. In addition to their extension and development of the calendar, they invented an ingenious mathematical system used in the count, based in a *vigesimal* numeral system, rather than the more familiar *decimal* system used in the majority of societies today.

This requires some explanation. A *decimal* system of numbers is one in Base 10, that is, in which the basic collection is ten numbers, after which a new unit is began. So, for example, we count one through ten, and then after ten we begin to count the *next* group of tens, eleven through twenty, and so on. The decimal system of numbers counts based on powers of ten. So we have 10, 20, 30, 40, and so on, beginning a new count of 10s for every ten. Although we are so familiar with this system as to completely take it for granted, there is no particular reason we *need* to count in a decimal system. It is believed that the reason many societies adopted a decimal system is the rather obvious one—we have ten fingers, and thus grouping numbers into units of tens is very intuitive when one is using fingers to count. But say we had eight fingers, or we just decided to adopt a different system. We could perfectly well adopt a Base Eight system of numerals, or any other Base for that matter. Base Eight would take each unit or place to have eight numerals, such that our counts would look like this:

1, 2, 3, 4, 5, 6, 7, 10, 11, 12, 13, 14, 15, 16, 17, 20, … and so on (Fig. 2.7).

The Maya mathematical system used for the Long Count was *vigesimal*, or Base-20. Thus there were 19 numerals before a new numeral place would be reached. The numeral writing system allowed this with much more ease than our own decimal system does (since we have fewer basic numerals than the Maya). The

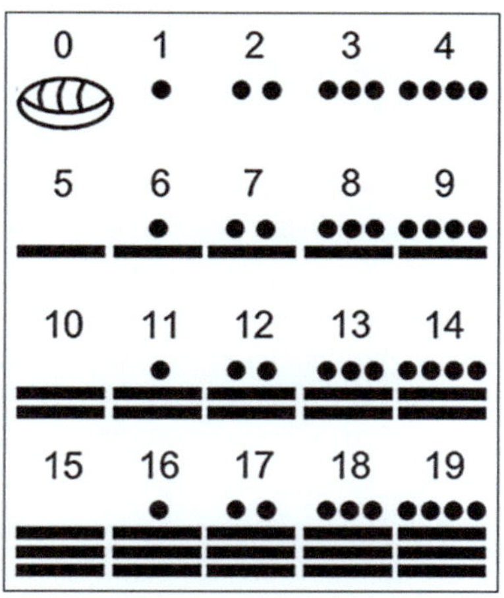

Fig. 2.7 The Maya numeral system was a Base-20, or *vigesimal* system, in comparison to our Base-10 (decimal) system. The system contained 19 distinct numerals, based on a *dot* signifying *one* to the *dash* signifying *five*, and a glyph of a shell, representing *zero*. The concept of zero was a necessary for the ability to shift places, as we do in our own decimal system. Without it, one has a clunky numerical system that is difficult to use for calculation, like that of the Romans (Photo credit: NASA)

Maya system used a dot to represent one, and a dash to represent five. Thus, three dashes and four dots would be the final numeral in the initial series before movement to a different place, with 20. The way this worked was similar to how our own place system works, but the Maya numbers were written vertically from top to bottom (similar to the classical Chinese), rather than left to right as in our own system, or right to left as in Semitic languages such as Arabic and Hebrew.

The bottom line is the 20s place, representing 1–19. The next line up represents the second 20s place (or 400th place), containing numbers 20–399. The next line up from there is the third 20s place (or 8000th place), containing numbers 400 (twenty 20s) to 7999 (where 8000 is twenty 400s). And so on, each place being a new multiplicand of 20. Thus, the numeral is equivalent to our numeral 1705, reading from the top line down thus: 1600 (four 400s) + 100 (five 20s) + 5. One possibility is that there may be numbers with *zero* units in any given place, and for this the Maya needed a conception of zero in order to make sense of this. Thus, we see the development in Maya (and earlier) mathematics of the concept of zero, which some scholars believed developed even before the discovery of the concept of zero in India, a region which is commonly accorded the credit for discovery of the concept.

These numbers were used in the unique and important "Long Count" calendar to keep track of days, months, and years, in a sequence beginning in what we know as 3114 BCE and continuing to 2012 CE.[18] The previous *baktun* sequence of the Maya Long Count calendar, in use during the Maya Classic Period, came to an end on a winter solstice day—December 21, 2012.

The Maya maintained two other calendars as well–the 260 day ritual calendar, or "short count" (*Tzolk'in*), and the 365 day solar year "vague" calendar (*Haab*).[19] While it may at first seem strange to us to have multiple calendars, when we think about what this really amounts to, as well as our own practices, we will see that it makes sense. Academic calendars are a good example, familiar to most of us—those who work in academia, as well as those of us who have gone to the university at some point. The academic calendar runs alongside the Gregorian calendar we use to determine years, but is not the same as this calendar, and has different beginning and ending dates, different holidays, etc. The academic calendar generally (for an institution running on semesters) has 9 months rather than 12 (as any academic who is paid on a 9 month contract is acutely aware), has years which begin not in January as do the Gregorian's, but instead in late August, and end not in December but in early or mid May. For those of us who live and move in academia, the academic calendar has as much, if not more, significance than the Gregorian calendar, which we also use. In my own life, then, I have at least two significant calendars, the academic and the Gregorian. It is easier for me to organize years of my own life in terms of the academic calendar than the Gregorian. When I think of

[18]The Long Count does not perfectly correspond to the vigesimal system, likely in part due to the fact that the relationship between days and years does not work out correctly. The 20 day *Winal* fits into the 360 day year (*Tun*) 18 times rather than 20. Higher counts of years follow the vigesimal system.

[19]Just as with the Long Count, the Maya did not invent these calendars, but adapted them.

2008 in the Gregorian calendar, for example, this is usually somewhat vague for me until I consider it in terms of two distinct academic years, the '07–'08 academic year, and the '08–'09 academic year. My memory of something that happened in 2008 can sometimes turn out to be something that happened in 2007, but was fixed in my mind to the '07–'08 academic year, which I associate with 2008. Most people who have at one time been students (which is just about all of us) thought in the same way while students. In addition, we see smaller scale calendars independently used. Our 7-day week count is independent from our month and year, even though we operate using all of these calendars. If today is a Wednesday, it may be Wednesday July 3 or Wednesday April 2. And July 3 may be in the year 1982 or the year 2015.[20] So we can see that use of multiple calendars is not foreign to us after all!

The Maya calendars, like our academic, Gregorian, and other calendars, played different roles in the community. The Long Count, 365 day solar calendar and the 260 day *Tzolk'in* calendar were often linked to one another (just as our academic and Gregorian calendars are linked). The operation of the 365 day calendar will be simple to anyone reading this, as our own calendar is a version of such a solar calendar. The Mesoamerican version did not include the conception of the "leap year" to calibrate the calendar every four years. The need for a leap year day, of course, arises from the fact that the full tropical year, which can be defined based on position of the sun in the sky, from solstice to the same solstice,[21] is not exactly 365 days, but 365.2422 days.[22] This means that every four years, a 365 day calendar will be a day behind the tropical year. With enough years passing without calibration, the calendar will slowly creep backward, and the seasons will diverge from the calendar. If we begin with December 21st marking the winter solstice, after 120 years, the calendar will be off by a full month, with January 20th marking the solstice (Notice also that since the discrepancy between the tropical year and solar calendar is not exactly 0.25, occasionally leap *seconds* have to be added to our calendar as well). It is unclear why the Maya 365 day calendar did not contain a calibration leap-year day or any other such device, even though the Maya were aware of the 1/4th day divergence between the year and the solar calendar. Perhaps the ritual integrity of the calendar, containing the same count of days each year, trumped whatever benefit may have been seen in including a calibration. As long as one knows and can keep track of the shift in the calendar, one can still track important dates such as the solstices, equinoxes, and zenith passages of the sun.

---

[20]Thanks to a reviewer for pointing out this additional fact about our calendars. Note that Wednesday, July 3, 2015 is an impossible date, on our calendars!

[21]Winter to winter or summer to summer. It could also be defined as the time between two of the same equinoxes.

[22]This is distinct from the *sidereal* year, which is based on the return of the earth to the same spot with reference to the background of the stars. The sidereal year is slightly different than the tropical year—it is 364.25636 days. This difference is due to precession of the equinoxes, which is discussed further below.

The 260 day ritual calendar is one of the most unique calendars of the Mesoamerican world, and an interesting and complex one. This was the calendar (along with the 365 day) used by the Aztecs, who did not adopt the Long Count so prized and perfected by the Maya. The 260 day calendar was broken into 20 named days (following the Maya vigesimal series of 20), starting with *Imix*, continuing through the series of 20, then starting with the next set. Along with each day, one of a set of 13 numerals was attached. Thus, each day of the calendar would be fixed with a day sign and a numeral, beginning with 1 *Imix*. The following day would be 2 *Ik* ('Ik' being the second day sign), and so forth, until the set of 20 days completed and returned to *Imix*, the 21st day. Because each day gets 1 of 13 numerals, the second appearance of *Imix* would not be 2 *Imix*, but 8 *Imix*. The next appearance of *Imix* after that would be 2 *Imix*, then 9 *Imix*, and so on, until every day sign had 13 rounds, after which the calendar would be completed, until the inauguration of the next ritual year.

There is some question of the reason for the establishment of this seemingly odd calendar. Why 260 days? There are a couple of possibilities here, both having to do with astronomical phenomena. One possibility is that the 260 day period corresponds to a distance between zenith passages of the sun. At the latitude of the southern Maya region, in which early developments in the formative Preclassic Period took place in cities such as Izapa and Kaminaljuyu,[23] the sun passes through the zenith on two days a year that are separated by 105 and 260 days respectively. Perhaps the calendar is meant to correspond to the latter period. But if so, why focus on this particular number, rather than the 105 day period? Another possibility has to do with an object we know had enormous significance in Maya culture and its predecessors: the planet Venus. The appearance of Venus as "morning star" is roughly 260 days (from its first appearance as morning star to its disappearance until its return as evening star).[24] However, this is not exact, as Venus is visible as the morning star for 258 days, and so it makes such a count curious as adopted by the astronomically highly proficient Maya (even if the calendar was created by a predecessor culture). Anthony Aveni (Aveni 2001) suggests that the calendar was meant to link the Venus appearance to the synodic period of the moon. The two will take longer to correlate than a single period of visibility as the morning star, and this follows the pattern of interlocking calendars popular in Mesoamerica. Whatever the reasoning behind it, the ritual calendar was an important part of Maya culture, and important dates are often given in both their ritual calendar, solar, and long count dates.

In addition to their interest in precision of dating, the Maya were also concerned with planetary motions, especially those of Venus. The Dresden Codex contains elaborate and detailed Venus tables, demonstrating the amazing accuracy with

[23]14.8° N. Kaminaljuyu is slightly south of this.

[24]A number of scholars link the *tzolk'in* to Venus' visibility, including Susan Milbrath (1999, 158). Floyd Lounsbury (1982, 163) proposes that the retrograde motion of Venus also held significance for the Maya.

which the Maya observed, charted, and predicted the motions of our neighboring planet. The question of course naturally arises–why were the Maya so interested in maintaining a detailed calendar and schedule for the risings, settings, and movements of Venus?

The first question may be easier to answer than the second. Any agriculture-based society, including our own, will have a need to determine with some degree of accuracy the seasonal changes. This requires having a calendar accurate enough to enable a society to determine when the warm and cold seasons will begin, in order to determine when to plant or harvest. But, we might ask, does this need necessitate a detailed and long-reaching calendar of the type maintained by the Maya? After all, a simple knowledge of which stars rise at sunset at a certain point of the year, such as the Pleades, will suffice for general agricultural purposes. And indeed we see across the globe that many pre-modern peoples used just such a system, with success, to determine times of planting and harvest. You will not gain any more or better crop yield if you determine the point of the year with greater specificity than these simple techniques allow. In addition, if agriculture is the primary concern, there is no need for a "long count" that will fix a given year in context of a count stretching multiple thousands of years. Our own calendar is an example of such a system. We can locate events thousands of years in the past or thousands of years in the future on our own calendar. We know that a supernova was visible in the northern hemisphere in 44 BCE, and that a total solar eclipse will be visible in North America on September 14, 2099, for example. There is no need for such an expansive calendar to aid in agriculture alone. What happens in 44 BCE or 2099 CE is irrelevant to my planting or harvesting crops. Indeed, what happened *last year* or what will happen *next* year is also irrelevant. And this is just why we find that many of the cultures that adopted simpler forms of determining planting and harvesting seasons and who were mainly interested in the calendar from an agricultural perspective did not tend to have calendars that tracked time beyond a year–the main (and only) relevant unit for purposes of agriculture.[25] This fact makes it difficult for contemporary historians to precisely date certain events recounted in the texts of such cultures.

In the case of the Aztec calendar, there is a related difficulty. This calendar has distinct days for a cycle of 52 years, at which point the cycle begins over again (unlike the Maya long count, which has distinct days for thousands of years). This makes it difficult to determine the dates in terms of our calendar, as we can know what year within a 52 year cycle an event happened, but not *which* 52 year cycle the event happened within. The 52 year cycles were not themselves organized into a longer system and ordered, so the third day of the second month of one cycle would be indistinguishable from the same day of a new cycle. Given the lifespan of people during the time of the use of this calendar, this makes practical sense. Hardly

[25]Even though the Maya, based within the tropics, did not have to worry about the issue of warm and cold seasons, as those of us in temperate regions do, they still took close note of the solar calendar, as it marked for them the critical wet and dry seasons, which, like the northern (and southern) warm and cold seasons, determined their periods of growing and harvesting.

anyone would live to see two instances of the calendrically identical day in two different cycles, so every day of one's lifetime would have a unique calendric marker, and one could refer to or fix events within a lifetime on the basis of this calendar. Of course, this calendar would make impossible a certain kind of deep history, in which events are precisely marked from past cycles. Should we conclude that cultures such as the Aztecs had no concern for history? Not necessarily. Our own contemporary way of thinking about history is not the only way humans can think about history, and the Aztec way represents one method of historical thought we certainly see elsewhere in the world as well.

Why, we might wonder, do we need an exact dating of certain events in terms of a contemporary calendar? A society may have a conception of history based in the succession of events, such that they may know that event *x* happened before event *y*, or even a conception of history in which the succession of events is not important at all. Indeed, we might put the question to our own conception of history–why is it important to know that some event happened in 44 BCE? What could 44 BCE mean to us more than "a really long time ago", given that we have no experience of anything even close to such a time? Having a context in which to place important events is key, of course, but this can be done in the absence of a detailed calendar tracking back thousands of years, even with such a calendar as the Aztec calendar. On many views of history, it is not important so much exactly when an event happened in connection with the current day, but its historical significance in general, consistent with it being something that happened "in days of old". Just *how* old becomes irrelevant after a certain amount of time.

So, given the compatibility of agricultural and historical aims with a less detailed calendar, there must have been specific, additional reasons behind the development and use of the Maya calendar, just as there are reasons behind the development and use of our own calendar. One strong possibility is that it was somehow due to the role of time and temporal context in Maya religion and philosophy, as well as due to the details and intricacies of Maya religious cosmology. Let's consider some of the features of Maya cosmology, before returning to think about time and the calendar more broadly.

Maya Cosmology

There is no doubt that the sky played a major role in Maya religion. The K'iche' Maya text *Popol Vuh*, written after the Spanish conquest in the southern region of contemporary Guatemala, recounts what purports to be a Maya creation story—an account of the creation of the world, leading down to humanity.[26] While there is likely Christian influence in the work (which seems obvious when one reads

[26]see Tedlock (1996)—this translation of the K'iche' original includes an introduction discussing influences.

through the first book, though the clear Christian parallels seem to taper off the deeper one gets into the text), it is also likely that this work preserves some much earlier features of Maya religion. Some of what we find here are likely beliefs that would have been held by the Maya of the Classic Period, during which the Maya intellectual project flourished, and the Maya people developed a complex language and textual tradition, formed a thriving empire, constructed majestic temple cities (even though they were never to become the kind of urban constructionists that their cultural predecessors in Teotihuacan were or the later Aztecs would become), and revolutionized astronomical observation and prediction.

In this text, we find a link between the sky and the other realms. The road to the underworld, *xibalba*, is said to be the dark rift apparent within the Milky Way galaxy, which trais across the night sky as a dim glow. One interesting feature of the night sky in tropical locations (among others) is that more of the Milky Way's path will be visible in the sky, and it will be higher in the sky as well. In more temperate regions further north, the Milky Way not only appears differently, but in the contemporary world we have the additional problem of rampant light pollution, all but blotting out our view of the path to *xibalba*. With modern cities and towns taking over our geography, even in some of the darkest sky sites available, the Milky Way cannot be seen from much of the United States or densely populated regions elsewhere in the world (Figs. 2.8a, b).

The Milky Way has an additional association in Maya thought—in addition to containing the path to the underworld, it is also associated with the "world tree" of creation.[27] It is unclear to what extent the K'iche' view of the Milky Way involving *xibalba* is shared by or comes from Classic Period Maya beliefs, but it does seem to be the case that the "world tree" view represents Classic Period Maya views, as it seems to be represented on inscriptions from urban constructions from the period.

As mentioned above in connection with the *tzolk'in* calendar, the planet Venus seems to have had enormous significance for the Maya. The Dresden Codex includes a Venus table, charting with startling accuracy the movement and the phases of Venus.

Many cultures have noted the significance of our sister planet Venus. It is, second only to the moon, one of the brightest objects in the night sky, and it is linked closely with the sun. In many cultures Venus is seen as both chasing and fleeing the sun. There are a number of interesting features of Venus that the naked eye astronomer will notice, features that are almost impossible to find in other planets, for a number of reasons. Venus, the second planet from the sun, is one of two other planets in our Solar System, along with Mercury, whose orbit is inside of that of the Earth. That is, Venus along with Mercury are closer to the Sun than is our own planet, and thus their orbits are both shorter and within the track of our own. This causes a few noticeable effects even for those observing the sky without the aid of a telescope. First, the movements of Venus (and Mercury) will appear to

[27]Friedel, Schele, and Parker (1993) argue for the association of the Milky Way with both the road to Xibalba and the "world tree" of the Maya.

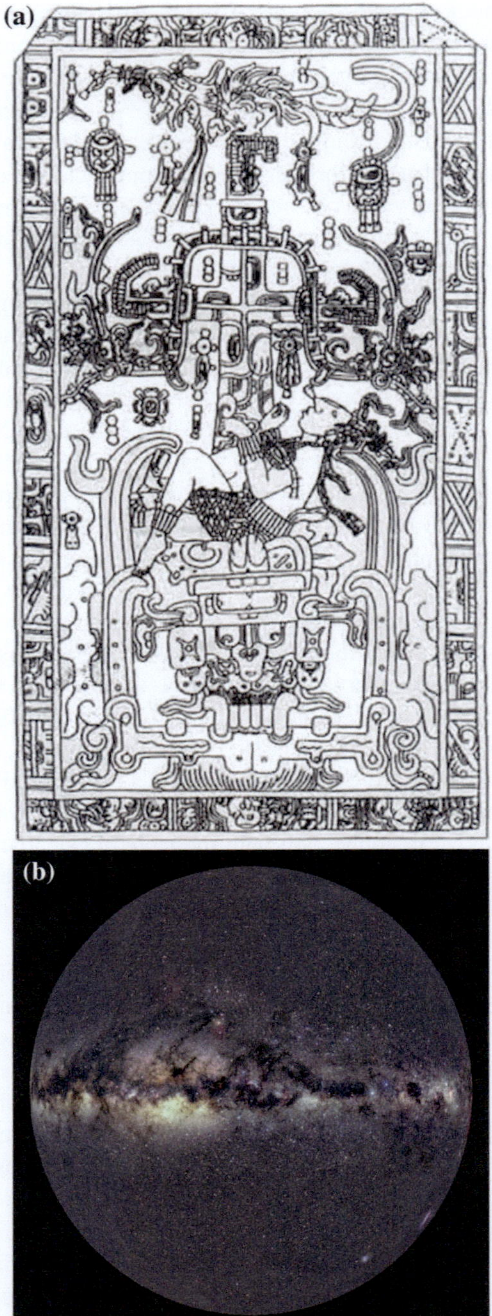
(a)
(b)

◀ **Fig. 2.8** The Maya associated the view of the Milky Way galactic plane across the night sky with both Xibalba, the realm of the dead, and also the World Tree, representing the cosmos as a whole. The first illustration is from the tomb of the ruler Janaab Pakal in Palenque, depicting the king's descent into Xibalba, from which the World Tree grows. The Milky Way was a familiar nighttime sight in the pre-Columbian world, just as were Venus and the Moon, but today there are few heavily inhabited places from which the band of the Milky Way is visible, due to rampant light pollution drowning it out (Photo credit: NASA)

us as closely linked to the sun. Neither planet will ever appear in the sky very far from the sun—although of course Venus can appear further from it than Mercury—and thus Venus only appears in the early evening or the early morning. It is from this that Venus gets its identity as both the "morning star" and the "evening star". The movement of Venus through its orbit causes this effect. When it appears on one side of the sun from us in its orbit, Venus appears in our night sky as an star, coming into view just after the setting of the sun in the west, as the glow of the sun fades. As it moves further in its orbit, passing us and (from our perspective) continuing around the sun (remember than Venus moves more quickly through its orbit than does the Earth, as all planets closer to the sun move faster through their orbits), Venus becomes invisible to us for a period of days when it is too close to the sun from our perspective to be seen. Then, as it moves to the other side of the sun, it once again becomes visible in our sky as a "morning star", trailing behind the sun and revealing itself just before the rising sun. It of course follows that Venus can only rise, that is, appear above the horizon, as a morning star, while as an evening star it makes its first appearance with the dwindling of sunlight, already above the horizon. This rising of Venus as morning star was a significant event for the ancient Maya.

Another feature of Venus due to its orbit within that of the earth is that, like the moon, Venus undergoes phases. Of the planets from our perspective on earth, only Venus and Mercury do this. The reason is that due to their interior orbits, we see these planets at different positions relative to the sun, just as we see the moon, and so we observe different parts of the planets lit up at any one time. Similarly to the moon, if we observe Venus from a 90° angle, we will see it in its "quarter" stage. There are though a couple of key differences between Venus phases and those of the moon. The moon orbits the earth directly, rather than the sun, so we are able to see all of its phases except for the new moon, when the sun renders it invisible. With Venus, we can never see a full phase, because when it would appear to us full, it is on the opposite side of the sun from us, which would make it visible only during the daytime, during which of course we cannot see it with the naked eye. Of course, there are today sophisticated astronomical tools that will allow one to observe Venus even in this phase, but absent such technology, it cannot be seen. Certainly no one in the Classic Period Maya world, or anywhere else before the advent of contemporary solar telescopes, would be able to see it (Fig. 2.9).

One interesting and perhaps counterintuitive feature of the phases of Venus is that Venus appears brightest in the sky not when it is its fuller phases, but instead in its less full phases. Why might this be? Remember that, because of Venus' interior

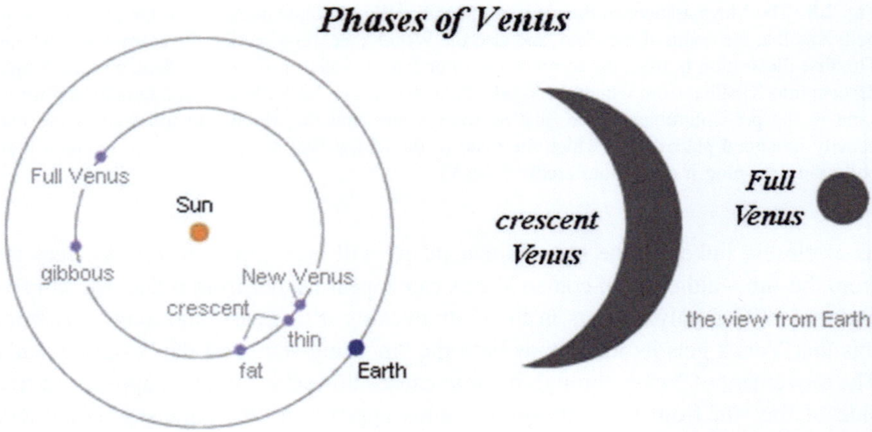

Fig. 2.9 Because its orbit is within that of the Earth (that is, it is closer to the Sun), Venus can be seen from earth to undergo *phases*, similar to the moon. Unlike the moon, however, Venus is brightest in apparent magnitude when it is in its crescent phase, rather than its full phase. The reason for this is that we only see Venus as full when it is directly opposite the sun from us. It is only in this position, however, when it is on the other side of the sun from the earth, which is the furthest distance from the earth possible for Venus. In fact, we never directly observe (with the naked eye at least) a completely full Venus, as the planet is too close to the sun from our vantagepoint to be visible in this phase. In its crescent phase, Venus is on the same side of the sun as the Earth, and as close to it as possible. Thus, while we see less of the full disk of Venus lit, its apparent magnitude is greater because of the proximity of this light (Photo credit: NASA)

orbit, it will be closest to the Earth when it is in its "new" phase (where Venus is between the sun and the Earth), and furthest from Earth in its "full" phase. This means that the closest visible pass to Earth by Venus is when it is in a waxing or waning crescent phase. It is just in this phase when we see the largest amount of area of Venus relative to the area on the celestial dome it takes up, rather than in terms of the amount of lit space on Venus from our perspective. This translates into a brighter appearance in phases in which less of Venus is visible. This is why, strangely enough, Venus is brighter as a crescent than it is when it is in the fullest state in which we can observe it, at which point it is closer to the opposite side of its orbit from the earth (we cannot, of course, observe Venus in its full phase with the naked eye, as its proximity to the sun from our perspective drowns out its light).

Venus' brightness, especially in those early phases, is remarkable, and people have been fascinated with it since ancient times. Indeed, in the modern world, where many people have lost the connection with the sky that humanity used to have, the brightness of Venus during especially bright phases has sometimes caused alarm. Authorities, news stations, and professional astronomers always claim (and often complain) that when Venus is bright they will receive a large number of calls asking the identity of that bright light in the sky, thinking it might be a stationary satellite, an experimental government aircraft, or something even more incredible.

The significance of Venus for the Maya was many-fold. It could be linked to their agricultural seasons in a certain way, and could also be linked to the power of the ruler, to understand and predict aspects of nature. The kings of the Maya world wielded enormous power, in part based on their abilities to predict, and thus control, the motions of heavenly bodies, especially the sun, moon, and Venus—the three brightest objects in the sky. In the Maya Classic Period, rulers of Maya city-states were also shamans and astronomers, and their authority in part derived from their ancestral connection to Venus.

In the Venus tables from the Dresden Codex, Venus is represented in monstrous guise, and illustrations are included beside the descriptions of the rising, phases, and setting of the planet. During the Classic Period, Venus was associated with one of the two culture heroes of the Maya world, Hun Ahaw and Yax Balam (called Hunahpu and Xblanque in the *Popol Vuh*). Venus was Hun Ahaw, a name that may be linked with a particular day on which Venus appears in the sky in the early texts. Hun Ahaw is also a calendar date, the first day (*hun*) of the month *Ahaw* (the word *ahaw* also means 'lord' in the Classic Maya language). The rising of Venus perhaps then signified the return to life of Hun Ahaw and his twin Yax Balam (represented by the moon), returning from the underworld *Xibalba*, where demons attempted to destroy them as they did their father, according to the *Popol Vuh*. Venus also had additional associations in the Maya world. It, like the sun, may have been linked to rulers and founders of lineages (Milbrath 1999, 196–197), as well as to the god of storms, *Chac*.[28]

Venus also had military significance for the Maya. It was seen as a patron or talisman of war, aiding the Maya in their battles. This perhaps can also be linked to the Hun Ahaw and Yax Balam myth. Venus, after disappearing, rises again as the morning star, and represents in a fundamental way the idea of rebirth and triumph at the heart of the story of the hero twins and their return to life and subsequent defeat of the Xibalbans. This would be powerful imagery and ideology in war, the Maya forces taking themselves to represent this power to overcome the darkness of the underworld and attain victory. This is a particularly interesting recent discovery, given the earlier view by some scholars that the Maya were a completely peaceful and sedentary people. It turns out that the Maya, like every other advanced civilization in the world, engaged in warfare quite regularly. And Venus (Hun Ahaw) was its patron (Fig. 2.10).

It is the Maya observation of Venus and their recognition of its patterns of motion that is the most impressive aspect of their astronomy concerning the planet. Determining the regular motions and orbital patterns of Venus is much more difficult than determining that of the sun or the moon, for example (although those were equally important). This is because, while the orbits of planets such as Venus are regular and elliptical around the sun, from our perspective on the Earth it is

[28]While *Chac* and *Tlaloc*, the Aztec god of storms, have commonly been identified as the same god, with *Tlaloc* simply the Aztec version of the god called *Chac* by the Maya, this has come into question by some scholars, including Milbrath (1999, 199–200) and Karl Taube (1992, 22).

Fig. 2.10 A number of different Maya glyphs referring to the planet Venus. One interesting (and confusing) feature of Maya glyphic writing is that a word could be written in many ways, often very different from one another, as seen in this drawing from "A Study of Maya Art" by Herbert Spinden, 1913

irregular, because we are not watching it from a static point. Rather, we are moving in our orbit outside that of Venus while it moves along it's orbit. Since Venus is closer to the Sun than the Earth, it moves more quickly in its orbit, and so passes us by as we go on our yearlong round of the Sun. This makes it much more difficult to discover the regularities in the apparent motion of Venus in our sky—though such regularities *can* be found, for those skilled and patient enough in observation. Fortunately, the Maya astronomers were both.

There was one celestial object even more important to the ancient Maya than Venus and the moon–the almighty Sun. The Sun, as giver of life, warmth, and light, is at the core of human existence, in every culture. Indeed, without the Sun our planet would not exist, and if the Sun were to disappear, our planet would go cold, unable to support life. The surface of the Earth would become a frozen rock, not very different from Pluto. Humans have always recognized the supreme importance of the Sun, and it has been revered in most civilizations throughout our history. We recognize that the patterns of the Sun's movement correspond to seasonal changes, to fertility and barrenness of the ground. When the sun reaches solstice at its highest point in the sky, the growth of crops are rampant (although one should not wait this late to plant them), and the Earth is green with plenty. When the sun hangs at solstice in its lowest point in the sky, in the winter, the earth is barren, cold, and dry. Or at least such is the case for those of us in the temperate zones outside the tropics in both the northern and southern hemispheres. For those in the tropics, such as the Maya, the sun has a different, although no less, significance. In the Maya world, the critical agricultural determination was not when the warm and cold seasons, summer and winter, would be, but instead when the rainy and dry seasons would take place. The key was to plant crops so as for their growth to coincide with the rainy season, which would then make the crop yield abundant. Then, harvest would take place before the onset of the dry season, to avoid the shriveling and dying of one's crops. Observation of the Sun played a critical role in this, just as it does in determining when to plant relative to warm and cold periods in temperate climates. Given that it is through observation of the Sun (along with that of the stars) that one can tell where one is in the year, then if one knows at what point in the year the rainy season begins, observation of the Sun (and stars) will help them determine when to plant crops, relative to this prediction.

The Sun had an additional significance for the Maya, a significance that the Sun had in many other cultures as well. Given the centrality of the Sun in human culture, there has been a tendency to draw a parallel between the importance and life-giving supremacy of the sun and that of the ruler. All across the world, in various ages, we see such an image projected, of the ruler as akin to the sun, or in some cases *identical* with the Sun. As we have seen in Roman culture, the sun was worshipped as a god, *Sol Invictus* "The Unconquered Sun", and during the Empire this god came to be associated with the Emperor. *Imperator*, the Latin term for 'emperor', was originally Augustus' attempt at giving himself a "humble" military title, basically equivalent to 'general'. The position of the Emperor began, then, adopting the symbolism and imagery of the humble servant of the people. With the decline of the Empire and the chaos of the various experiments in rulership, the imperial image began to change, and with the unification and rise to power of Constantine, the identification of the Emperor with a god, in this case the Sun, became explicit. Even after his conversion to Christianity, Constantine represented himself on coins and elsewhere in the guise of *Sol Invictus*.

Maya kings were also associated with the sun. There are many illustrations and stone inscriptions at Maya urban sites that identify kings with the sun. The title *K'inich Ajaw* ("Sun Eyed Lord", or "Radiant Sun Lord"), associating the king with the Sun or the Sun God, was often applied to rulers.[29] At the recently uncovered Temple of the Night Sun at the site of El Zotz in Guatemala, there is sun imagery linked to the person of the ruler.

The ruler, associated with and seen as having a power over the movements of the Sun, and gaining his own power from the Sun, naturally would have taken it as important to both have accurate skywatchers charting the movement of the Sun, from solstice to solstice through the year, and also to build monuments commemorating the Sun, which was also himself. Perhaps the grandest of these Sun temples is in the late Maya-Toltec city of Chichen Itza in the northern part of the Yucatan peninsula in modern day Mexico. The pyramid of Kukulkan (identical to the better known Aztec god Quetzalcoatl, "plumed serpent", whom, so the story goes, the Aztecs mistook the Spanish conquistador Hernan Cortes for when he first appeared with his fleet off the coast of the island city Tenochtitlan), is today referred to in Spanish as the *Castillo* (castle). While solstices and equinoxes are often

[29]Colas 2003 discusses *k'inich* as Sun God as well as the link between the Sun and the ruler in Classic Maya imagery and texts.

marked by sunwatching cultures, and these passages of the sun are important events in the annual calendar, those within the tropics experience an additional phenomenon of the sun that those outside the tropics in the north and south do not. At exactly two points in the year for equatorial observers, the Sun will pass through the zenith. The Sun will on each of these days mark noon directly overhead, at the center point of the dome of the sky. In temperate regions, such as the entire United States, Europe, and most of Asia, the sun never reaches the zenith, and the further north you go, the lower in the sky is the sun's highest point, at the summer solstice.

The Kukulkan pyramid was aligned to the dates of the Sun's zenith passage. At noon on the days of zenith passage, when the Sun hangs directly overhead at the summit of the sky, its light shines in a direct line, or at a right angle of 90° with the ground. One interesting result of this is that during this noontime, objects will cast no shadows, being completely bathed in direct light from above. On the pyramid itself, the construction is such that at the equinoxes, a row of shadows is created by the effect that processes down the stairwells of the pyramid, linking the top to sculptures of the head of the plumed serpent god at the bottom of the pyramid. This creates the magnificent effect of a serpent descending from the top of they pyramid-Kukulkan descending from his throne. The descending serpents point in the direction of the sacred cenote at the site, a massive sinkhole of the type common in the Yucatan, which collected essential water in the dry region.[30] Elsewhere in the Chichen Itza site, narrow pillars set up for the purpose mark the zenith as the perpetual shadows they cast disappear in the noontime sun. This was clearly an event of central importance at Chichen Itza and elsewhere, and there is much more attention paid to this marker point in the sun's yearly journey than there is to either of the solstices. The site of the Caracol in Chichen Itza has been determined to be an astronomical tool, the use of which enabled the Chichen Itza astronomers to accurately track the positions of the Sun, the Moon, and Venus especially, but ultimately any celestial event they wished to (Fig. 2.11).

We find throughout the Mesoamerican world, in Maya sites as well as others, of the Aztec, and the earlier Olmec, the construction of sites, sacred as well as secular (although in the ancient world here and elsewhere, the two were not distinguished so starkly as we tend to make them today), that are oriented to some important direction based on the sky–whether this is the simple cardinal points orientation that we find in sites like the Cahokia mounds in Illinois, or a more elaborate orientation to some specific celestial event, such as at Teotihuacan, or at the Anasazi sites at Chaco Canyon in current day New Mexico. In the culture of the Aztecs, one of the inheritors of many of the features of Maya learning, the Sun had enormous significance, perhaps even greater than it did for the Maya. The Aztecs seem to have had a concern about the constancy of the Sun, seeing it as a god that needed to be propitiated in order to consistently perform its life giving role. They believed that it was necessary to make human sacrifices to the sun, in order to please it and to

[30]Thus the imagery of Kukulkan descending may signify the coming of the rainy season in May, which corresponds with the May zenith.

Fig. 2.11 The pyramid of Kukulkan in Chichen Itza (in the Mexican Yucatan Peninsula) is often called 'El Castillo' today, the Spanish name given to it during the modern period. The pyramid was designed so that on the equinox days, the shadow created by the corners of the pyramid at sunset would fall on the staircases. The shadow made was meant to represent the descent from heaven of Kukulkan (the "plumed serpent" god known to the Aztecs as Quetzalcoatl). In this photo, the head of Kukulkan can be seen at the base of the pyramid, linked to the staircase and the equinoctial shadow forming the serpent body of Kukulkan, descending from the *top* of the pyramid

ensure its continued rise each day. The great extent of Aztec human sacrifice is well known—although the Maya also performed human sacrifice, as did other cultures in central and South America, including the Inca in the Andean region of South America.

The Maya had a similar practice of human sacrifice, although the Maya sacrificial rite is not linked directly to the sun as is the Aztec rite. The Aztec sacrificial ceremony involved a priest cutting the heart out of a still living person, and offering it in a ritual to the sun. Archaeologists have found both Aztec and Maya ceremonial stone vessels that held the hearts of sacrificial victims. The *living* heart was particularly pleasing to the sun, and it was for this reason that the victim's heart would be removed while alive, rather than killing the victim first and then removing the heart.

Human sacrifice is by far the most controversial and startling aspect of the pre-Columbian cultures of Mesoamerica. Some have tried to soften the horror of human sacrifice by pointing out that victims often saw sacrifice as a great honor, their deaths ensuring them the enjoyment of some sense of immortality. And while this is true in some cases, it is not the whole story. The Maya and Aztecs both tended to sacrifice not important people in society, which one might expect if they truly believed it to be an unqualified honor, but prisoners and captives in battle. While there are exceptions to this, it may be the case that rival nobility, rather than a group or city's own, were sacrificed. One unique feature of warfare in Mesoamerica is that very often battles were not waged in order to produce or maximize casualties on the other side, but primarily in order to attain prisoners, who would then be used for the human sacrifice that ensure the continued patronage of the sun. These prisoners would hardly have seen it as a good thing to have their hearts cut out ceremonially on a platform of the pyramid of a rival city. There was, however, a relationship of mutual benefit between any two cities. While one belligerent sought to take prisoners from the other for sacrificial purposes, the other side equally sought such prisoners. It was this mutual need that ensured that warfare remained primarily aimed at capture rather than destruction (and perhaps that left the Aztecs unprepared to deal with the armies of the very much destructive-minded Spanish in the 16th century).

There are a number of theories as to why the Mesoamerican cultures adopted human sacrifice, which range from population control to military reasons. There is also, of course, the reason that they perhaps actually believed in their religion, and thought that sacrifice was necessary to ensure the proper working of the cosmos, however it had first been practiced. It is unclear how and why they would have noticed a link between the sacrifice of a human and the continuation of the world, but it seems clear that this became the belief.

As mentioned before, the Sun played a central role in the religious apparatus of human sacrifice, and indeed the Sun was a somewhat fickle and ambivalent life-giver that, in the Maya context, could easily take life away as easily as it could give it. While we in the temperate zones of the northern hemisphere, in the modern world in which we are not dependent on local agriculture, might have a hard time understanding why a people might think the Sun needs to be propitiated, it makes more sense when we consider the features of the Maya world. First, in their dependence on local agriculture (they could not have food air-shipped from thousands of miles away), conditions of drought would have enormous effect on their ability to produce food. Second, given the seasonal patterns of the region, they did not experience a warm summer and cold winter, as people in the temperate regions do, but instead experienced a rainy season and a dry season. The dry season was ruled by the Sun, which parched the land and made it infertile. Thus the Sun could sometimes be the giver of life, and sometimes the destroyer of life. It was important to the Maya (as well as other Mesoamerican peoples) to ensure that the Sun continued to bring benefit to the people and avoided destroying them. And it was, in part, human sacrifice, that made this possible.

1054 for the Maya: Some Speculations

Did the supernova of 1054 have any impact in the Maya area? Felix Verbelen proposes that the scribes of Chichen Itza recorded the supernova of 1054 in the Venus Tables of the Dresden Codex, as the supernova may have appeared as a "second Venus" in the sky over the Yucatan.[31] While Verbelen's conclusion is controversial and relies on calendar correlations that are not generally accepted by most scholars,[32] we might imagine that the 1054 supernova would have been an event of significance for any skywatcher concerned with Venus. Whether or not the Dresden Codex records the supernova, there are still interesting questions surrounding the event. Did the Toltec-Maya of Chichen Itza find any significance in the cataclysmic celestial event hidden to us today, which they clearly must have observed? Perhaps the supernova coincided, whether purely coincidentally, or purposefully, with some major event in Chichen Itza society. If so, what might such an event have been? It is very likely that if some important event was correlated to the 1054 supernova, it would have been a military conquest of some kind. Venus, as we have seen above, was the taken seriously as the patron of war, and its position often had a hand in determining the timings of invasions and other military maneuvers. If the record of the 1054 supernova was contained in the Venus table, it too likely would have been connected to some militarily significant event.

What could that event have been? There are a number of possibilities, but perhaps two that are most interesting. Yaxuna and Coba, two Maya cities slightly to the south of Chichen Itza in the Yucatan,[33] went into decline in the late 10th and early 11th centuries. It could be that Chichen Itza had a role in conquest of those cities—perhaps the people of Chichen Itza had finally overcome these cities decisively following the 1054 supernova, and this had been seen as a significant omen worthy of memorialization. An even more wildly speculative but interesting possibility is that the supernova corresponded with connections between the Toltec-Maya and their neighbors not to the south, but rather to the *north*, across the Gulf of Mexico, in North America. Today archaeologists note the startling similarities between the cultural renaissance in North America, such as that of the Mississippians, and the cultures of Mesoamerica, including the Maya and Aztec.[34]

---

[31]Verbelen 2000 proposes that the 1054 supernova corresponds to a date given in the Venus tables of the Dresden Codex, but his proposed date of May 10, 1054, conflicts with archaeological evidence for the date. Especially since the *tzolk'in* date, 4 Ajaw 8 Cumku, was an important *tzolk'in* date that inaugurated new periods. (Kelley and Milone, 2005).

[32]The most widely accepted calendar correlation is the Goodman-Martinez-Thompson (GMT) correlation. Verbelen is careful to mention this discrepancy at the beginning of his article, and maintains that his reading may be falsified by further evidence.

[33]Yaxuna is closer and almost directly south of Chichen Itza, whereas Coba is further to the southeast, closer to the Atlantic coast of the peninsula.

[34]A number of archaeologists, including Timothy Pauketat of the University of Illinois, who works on Mississippian culture, and Gerardo Gutierrez, Mesoamerican specialist, and Stephen Lekson, Southwest specialist, both of the University of Colorado, argue for a robust connection between

As we will see below, this new flowering of culture in the north, so similar in form to the culture of the Maya, begins seemingly as if from nowhere, in the years surrounding 1050 CE.

Mississippian Culture

One cannot help but have a visceral and profound feeling of the pull of time when standing atop one of the pyramids of the Maya, or the ceremonial mounds of the Mississippian peoples of North America. In addition to this, there is the additional sense of deep mystery when one enters the sites of the mound building peoples of North America. Who *were* these ancient people, one wonders. What was the significance of what they built? What were their beliefs, their relationships, and their understanding of the sky? While all of these questions will of necessity be much harder to answer than they are for cultures such as the early Maya, they can be answered to some extent.

Why are these questions more difficult when we come to ask them of the Mississippian peoples? There are two major reasons for this. First, the Mississippian cultures, although they built ceremonial centers, pyramids, and majestic cities every bit as large in scale as those of the Maya (and Aztecs), they did not build with stone, but instead with earthen materials, which are naturally less long-lasting than stone. In addition, one cannot carve inscriptions on the earth, or at least inscriptions that will last much longer than one season. Thus, we do not find the elaborate inscriptions that we do in the Maya world, which tell us a great deal about Maya culture, history, and astronomy. Secondly, the Mississippian peoples appear not to have developed a *textual* culture. There was no system of writing that we know of, and thus no texts. As mentioned above, in the Maya case, there are only four texts that we know of available today because of the routine destruction of Maya texts by the Spanish–and it is impossible to avoid reflecting on what kind of amazing discoveries concerning the Maya could be made if only we have access to the many other texts that have been lost. But in the case of the Mississippian culture, we have no texts at all to give us a description or even clues as to how people lived, thought, or understood their world. Without texts, writing, and distinct architecture, then, we have to rely on other materials and things to try to gain some sense of the world of the Mississippian peoples.

Studying sites such as Cahokia and numerous smaller sites in the region, both in their orientation as well as the artifacts that can be found there, is one way to approach the problem of answering who these people were, and most importantly for our purposes here, how they understood the sky. Another way is to look to the

(Footnote 34 continued)

North American and Mesoamerican cultures in the early years of the 11th century, coinciding with the beginnings of Mississippian culture and those of the Southwest including the Anasazi.

historical and contemporary peoples who are the descendants of the Mississippian peoples. When we find broad similarities among a number of different descendent groups, this may be some indication that these aspects of culture trace back to the Mississippian peoples like those who lived in the city of Cahokia.

The astronomical culture of the Cahokians in many ways must have been similar to that of the Maya. However, there were a number of key differences in astronomy of North America, and of Missisippian culture in particular. As with the Maya (and many other cultures throughout the globe), time-keeping was a major consideration of the astronomy of north American cultures. There was also likely religious and governmental significance of the sky. Finally, there is a great deal of evidence that the Mississippian peoples were, just as the Mesoamerican cultures, close watchers of the sky. Indeed, there is some reason to believe that the flowering of Mississippian culture was influenced by Mesoamerican culture. Astronomy was central in the lives of this people, just as it was for the Maya (as well as a number of other things, including human sacrifice–although this never reached the extent in the Mississippian region that it did in Mesoamerica). At numerous sites throughout the wider region of the Mississippian peoples, observatories can be found at central locations, suggesting that not only did these people watch the sky, but that it had a profound and central religious significance for them. In our own contemporary scientific culture, although we are able to watch and understand the sky with much more precision and accuracy in certain respects than people in the days of Cahokia due to our modern technology, astronomy occupies nowhere near the significance to us that it did to our ancestors anywhere in the world, but particularly in North America.

On the southern edge of the city of Dayton, Ohio rests an archaeological site on the west bank of the Great Miami River. This site was, archaeologists believe, occupied by a prehistoric group referred to as the "Fort Ancient" people- once thought to be a branch of the wider Mississippian culture, but now generally accepted as descendants of earlier Woodland peoples. The Fort Ancient culture was named after the much larger site about 20 miles southeast of this village, along the Little Miami River in Warren County, Ohio (we don't know what they would have called themselves). The Fort Ancient site itself was a ceremonial earthwork high above a steep gorge, and was likely built by a distinct and earlier people, today called the Hopewell (also named after a site near Chillicothe, Ohio).

This village in Dayton was constructed in an interesting way. It consisted of thatch and daub huts, built in a circle around a central point, in which there would have been a large pole marker. Archaeologists who have uncovered and worked on the site showed that this site was not only a settlement, but was also used as an observatory. And in a particularly ingenious display of elegance and efficiency, the huts themselves serve astronomical purposes in addition to their purpose as homes. The village itself is an observatory. The site was named "SunWatch Indian Village" in recognition of this purpose, and archaeologists in collaboration with the local natural history museum Ohio are reconstructing the site, which has been opened as a museum and park.

One of the oldest and simplest forms of positioning objects in the sky, especially on the horizon, is to use *sight lines* along the ground. We have already encountered this practice, in both the alignment system of construction of Teotihuacan, and the Caracol observatory in Chichen Itza, for example. There are many other ancient observatories that work on this principle, including Stonehenge in southern England, and let's not forget the ancient city of Cahokia, to which we will soon return.

There is an additional feature in use at the SunWatch observatory–one just as old and revered as the method of using sight-lines: use of the *gnomon*. The central pole in the village consisted likely of a tall wooden pole, at least 50 or so feet high (the higher the better, for gnomonic purposes). Because the sun occupies different positions in the sky at different points in the day and also at different times of the year, the shadow that this large central pole casts will change. This is the basic idea behind the sundial, an idea we will discuss at length in chapter three. Because of this feature, one can use a gnomon to construct a kind of clock. When the sun rises in the east, the pole will cast a long shadow to the west, as the sun is shining directly onto the eastern-facing side of the pole. As the sun rises in the sky, the shadow will slowly move to the northeast, finally pointing north at noon when the sun is at its highest in the sky, then gradually moving to the southeast as the sun sets, before ending in exactly the opposite position from its position at sunrise, as the sun sinks beneath the horizon. Knowing this, one can roughly (or more accurately, with some additional knowledge and tweaks) tell the time at any given point in the day. We know that when the gnomon's shadow is cast directly north, it is noon. If we are near one of the equinoxes, we know that when the shadow is at 45° northeast, it is 3 o'clock pm. There are of course some complications, because the path of the sun through the sky does not remain constant year-round, of course (Fig. 2.12).

And this fact leads to an additional use of the gnomon. One can use the gnomon to determine the solstices. At the winter solstice (in December in the northern hemisphere, June in the southern), the sun will reach its lowest extent in the noontime sky, and thus the shadow cast by the gnomon at noon will be longest at this point than at any other time of the year. If one has marked, for any given gnomon shadow, the point at which the shadow reaches its longest extent (a solstice line), then one can determine the winter solstice by noting when the shadow reaches this line at noontime. The same goes for the summer solstice, at which the gnomon's noontime shadow will reach its lowest extent, corresponding with the sun's highest ascent. Of course, all of this holds only for astronomy in the northern or southern hemisphere. The situation is much more complicated in the tropics, in which, as we've seen, the sun crosses over the zenith. There will be two days of the year that a gnomon will cast no shadow at all, just as we see with the pillars in the Temple of the Warriors in Chichen Itza, and because of the crossing of the zenith, shadows will be on different *sides* of the gnomon at different parts of the year. In addition, gnomon shadows will never register very much seasonal change in the tropics, as the sun never shifts very far from is overhead path. And due to the shift of the sun across the zenith, this will also make using the gnomon as a timekeeper

Fig. 2.12 A model of the Fort Ancient village today called SunWatch (Dayton, Ohio), as it may have looked in the 13th century CE. The central pillar was used as a gnomon and as a sighting point, on the same principle as the "Woodhenge" site of Cahokia. The entire layout of the village served as a calendar, on which significant risings and settings along the horizon could be marked, using the central pillar and homes as markers. (Photo credit: Dayton Convention and Visitors Bureau)

much more difficult. It was likely for these reasons that the gnomon was a much more central feature of North American astronomy than it was of Mesoamerican astronomy.

In addition to its gnomonic uses, the central pole also marked the position from which an observer would use sight lines to determine the positions of rising and setting of certain important celestial objects, such as the sun and moon, Venus and other planets, and the Pleades star cluster. Once the cardinal directions are found and marked (north, south, east, and west), one can continue on to add additional points between the cardinal directions, in a circle (there are also other ways to do this as well so as to increase precision, but this is the way things were done in most ancient sites as well as at SunWatch and Cahokia). In particular, one would want to mark the position of certain important risings or settings, such as the rising or setting point of certain stars corresponding to particular points of the year. Most ancient cultures, and most cultures in the Americas as well, used the Pleades star cluster in this way, to determine when to plant crops and when to harvest.

With the central pole as the observing point, and a sufficient number of posts or other markers around the circular perimeter, one can determine the position of some celestial object or event relative to the apparatus of the observatory. If a group knows already in what direction certain important events will happen, such as sunrise or sunset on important days like the equinoxes or the solstices, one can build posts, huts, or other markers so as to line up with this event when viewed from the central post. At SunWatch, the huts making up the perimeter surrounding the central post were built in just this way—as the sun rises on the morning of the equinox, for example, it shines directly through an intentionally placed gap between two of the huts. The genius of SunWatch is that the huts themselves play the role of posts, marking the cardinal points as well as the rest of the circle, and marking in various ways the important annual events.

This may have been knowledge transmitted by the culture of the builders of Cahokia. There were connections between Fort Ancient and Mississippian cultures for many years, especially the years between 1300 and 1500 CE, during which the SunWatch site was constructed.[35] At Cahokia there was a similar construction, labeled "Woodhenge" after the European site of the same name, but where the role of peripheral markers was not played by houses or huts, but by wooden pole markers, smaller than that of the central point. Cahokia's Woodhenge was devoted completely to astronomy, located outside of the heart of the city, including the ruler's palace on the largest earthen pyramid, today referred to as "Monk's Mound". Perhaps because of the massive size of the city, in comparison with SunWatch, the Cahokians could afford to build and maintain an observatory purely devoted to astronomy. There was also an additional gnomon pole in the central area of the city ("downtown" Cahokia, if you will) that likely served as a religious-ceremonial point, aligned with Monk's Mound, while at the SunWatch site, the central pole plays both roles. Necessity is the mother of invention, of course, and the people at SunWatch found an ingenious way of realizing multiple functions—astronomical observation, living areas, and religious center—with one single complex. SunWatch shows us an excellent local and consolidated expression of the astronomical knowledge of the Mississippian peoples as found in their metropolis and cultural center, Cahokia.

The city of "New Cahokia", as archaeologists refer to it, was built over top of an older city, referred to (creatively) as "Old Cahokia". Old Cahokia was a village of the type one may have seen many of in the region, large for the area but not magnificently so. Old Cahokia was home to around 1000 people, and did not have massive mound structures or observatories as we see in the new city. Around 1050 CE, the city of New Cahokia was built, and it was envisioned on a massive scale. This was not a city that slowly grew into a major metropolis. It was *intended* to be such, from the moment it was built. As remarked above, some have believed there was a connection between the 1054 supernova and the new city of Cahokia. While there is no direct evidence to establish this, there is certainly some plausibility to the view, especially when we note the new features of Mississippian culture that seem to become dominant with the rise of New Cahokia. The rise of enormous mounds like Monk's Mound at the northern edge of "downtown" Cahokia, as well as the devoted "Woodhenge" observatory site and the central city gnomon all seem to

[35]Scholars have shown that use of certain forms of pottery in Mississippian culture were adopted by Fort Ancient peoples during this period (Cook and Fargher 2008).

suggest that astronomical observation was a major aspect of this new city and culture born at Cahokia. The rise of New Cahokia is the dawn of Mississippian culture, and these central astronomical aspects of Cahokian culture are shared elsewhere in the region, including later sites like SunWatch built by people likely influenced by the Mississippian culture. Was it an astronomical event that sparked this shift? 1050 CE is suspiciously close to 1054. Might it have been an amazing supernova that caused this shift?[36]

We thought about Mesoamerican astronomy above, and its unique features. Some experts believe today that there may have been connections between the Mesoamerican cultures and those of North America like the Mississippian. There are undeniable and startling similarities between the cultures of the Maya and Aztecs and northern cultures like the Mississippians and the Pueblo. Archaeologist Timothy Pauketat writes:

> There are strong suggestions that the Cahokians, in building their vision into the landscape, drew on Mesoamerican models. Their possible descendants or those of their allies or enemies practiced Mesoamerican-style human sacrifice, incorporated obelisklike posts into their worship, relayed stories of superhuman men and women who wore distinctive garments and ear ornaments, used Mesoamerican-type flint daggers, and understood the cosmos in ways occasionally parallel to Mesoamerican notions. (Pauketat 2009, 7).

Thinking of their astronomy in particular, the new developments we see at Cahokia seem startlingly similar to those of the Mesoamericans, especially the Maya. The building of pyramids as both ceremonial altars as well as platforms for astronomical observation, aligned with important risings and settings along the horizon, was a major feature of both Mississippian culture and Mesoamerican cultures. The largest such structure at Cahokia, today called Monk's Mound (after the French Trappist monks who owned the land in the 19th century, lived near it and grew crops on top of the large mound), played both of these roles, and also served as the place of residence for the leader of the Cahokian community. The ruler's house was built atop the mound, and it was from there he and presumably his astronomers as well, could completely see the horizon, above the trees. The problem of how fully observe the horizon was an important one for cultures in places where trees and other vegetation were thick. In the Maya world, astronomical interest and civilization first developed in the region in which it was most difficult to see the horizon, as it was most densely populated by plant life. In the central highlands of modern day Guatemala sit the ruins of the ancient Maya city of Tikal—a city that thrived during the Classic Period, and whose political rise is associated with the beginning of the Classic Period and the Maya renaissance. As any visitor to the region around Tikal will understand, it will quickly become an issue to figure out how to get any view of the horizon. Tikal is in the middle of a tropical rainforest, with enormous trees and thick vegetation. There are really only

---

[36]A number of scholars have considered the possibility of the 1054 supernova's causal role in the construction of New Cahokia, Cahokia's "Big Bang". Timothy Pauketat discusses this position in (Pauketat 2009).

two options in such a region—either try to clear as much of the forest as you can, so as to have the trees so distant from your observing point that they do not obscure your view of the horizon, or instead build *upward*, creating structures to try to get above the treeline, and thus observe the total horizon from there. Although the former option is theoretically possible, the labor involved would be truly immense, and it would be of more use to turn any such cleared fields into agricultural area. It is much less labor intensive comparatively, and uses much less space (so space can be devoted to other important things like agriculture) to build a structure that can get one atop the treeline for observation. This, combined with a somewhat cleared area, will give one a much better view of the horizon.

This is just what the people of Tikal did. And one can still today see the pyramid tops with their platforms jutting out atop a sea of tropical forest (though the forest almost certainly did not encroach on the city during its period of use as it does today). It is an amazing sight, and shows both the ingenuity of the builders of these structures, as well as their commitment to astronomy, and its deep significance to their lives. At more northerly cities like Chichen Itza in the Yucatan Peninsula, thick forests were not a problem. The Yucatan is a uniquely excellent place for following horizon astronomy. It is flat, dry, and thus sparsely populated with plant life. Of course, these same features make it much more difficult to thrive and survive than it is further south in the rainforest. It may have been in part for this reason that the Toltecs, who originated further west in the Valley of Mexico, were able to conquer the Yucatan region and thrive in this environment. They knew how to make the most of what such land offered.

The pyramids at Cahokia were not exactly the same as those as that one finds at Tikal or Chichen Itza. Because they were earthen structures, they do not today seem as majestic and imposing as the Maya or Teotihuacan structures, which are older than the Cahokia mounds but have lasted longer because of the nature of the material with which they were constructed. The mounds at Cahokia are still impressive, no doubt, but they do not quite invoke the overwhelming awe one experiences when seeing the pyramids of Tikal or Teotihuacan. They must have been much the same in their day. As I mentioned above, one can walk to the top of Monk's Mound, and from there you get a sense of how truly massive it is. Ascending other pyramids in the Mississippian region is very similar. They are often very steep, and appear larger than they look from the ground.

One interesting feature of Monk's Mound is that it was the quintessential "multi-purpose" structure. In addition to the features I mentioned above, being the home of the ruler, astronomical observation point, and ceremonial center, it had also originally been a burial mound. There have been remains found toward the base of Monk's Mound, and it seems that it was used in the standard burial mound sense that we find throughout North America far before the creation of New Cahokia. It is possible that Monk's Mound began its existence as a burial mound in the older community that inhabited Cahokia, and when New Cahokia was built, it was re-purposed and renovated. It seems to be the case that it continued to be used as a burial mound on occasion, and it may have been rulers and relatives of rulers who were buried here after the construction of New Cahokia.

There seems to be an interesting feature of burial mounds across the Mississippian region that corresponds to one of the uses of Monk's Mound, and can help make sense of the combination of the burial and ritual features. Many ancient burial mounds seem to be built on relatively high ground, and thus perfect for astronomical observation of the horizon. In the town of Miamisburg, Ohio sits the well-known Adena culture (800 BCE–100 CE) site referred to as the Miamisburg Mound. The mound itself, a 70 foot high conical mound that is the tallest such mound in the state, and one of the tallest in the country, rises from a point that is already the highest point in the city of Miamisburg, and one of the highest points in the entire valley of the Great Miami River, atop a hill jutting out from the countryside 100 feet above the valley. Walking to the top of the Mound, one can see over fifteen miles in the distance in every direction. It is easy to see downtown Dayton about 12 miles to the northeast, and far beyond, and the astronomical advantages of the site from the top of the mound are stark and obvious to those climb to the top (Figs. 2.13a, b).

Miamisburg Mound was first excavated (very sloppily) in 1869, and there were a number of burials found at the higher levels of the mound, but none further below. An earlier age of archaeology was not as careful with artifacts as is this one. While one may expect this because of the relative newness of the higher layers, such that the bodies have had less time to decompose, one very interesting feature was found near the bottom of the mound. There was a hollow area, some kind of chamber or room. This could have been the burial site for an important leader, or could have been an older ritual chamber, or meeting site. I think it is likely, and would be consistent with cultural features we find in later Mississippian culture such as that at Cahokia, that the Miamisburg Mound was also used as an astronomical observatory and ritual center. Another very interesting feature of the mound, particularly in comparison to Mesoamerican constructions, is that at one layer of the mound, a stone surface has been discovered—about 25 feet deep into the present mound. This means that at one time, people in the region decided to face the mound with stone, thus creating a lasting monument more akin to those in the Mesoamerican region. This also shows that the peoples of North America understood that stone could be used to construct ritual or living centers, but that they preferred using earth and wood, more (relatively) temporary materials. This was shown definitively in the mound itself, which was covered back over in successive layering with earth and wood. Earth and wood material for building would have, of course, been preferable for a people who had reason to think they might be on the move, and we do seem to see this much more in North America than in Mesoamerica. The weather patterns in the north are much more harsh, with longer and colder winters, and people in the north would likely have to move in search of food in ways the Mesoamericans never did. For whatever reason, the Adena peoples and the Mississippians after them decided that earthen materials were the way to go—even the necessarily sedentary peoples of New Cahokia.

But the Miamisburg Mound site predates the birth of the Mississippian culture at New Cahokia, by about one to two thousand years. The mound was built by a people referred to today as the Adena culture, named after the estate of the once

(a)
(b)

◀ **Fig. 2.13** The Miamisburg Mound, in Miamisburg, OH along the Great Miami River, was built by people of the Adena culture during the mid first millennium BCE. It was not constructed at once into its current form, but was built up in layers over many hundreds of years as a burial mound. It may have begun as a ceremonial center. The vantagepoint of the top of the mound offers the best natural view of the horizon available anywhere in this section of the Miami Valley region. It seems plausible that it would have been used for astronomical observations at some point. Photos by the author

Governor of Ohio, Thomas Worthington, in Chillicothe, OH on which a prominent mound was discovered and excavated. The Adena culture presents a mystery of its own, as it abruptly seems to disappear around 400 CE. Mounds such as Miamisburg which have been excavated have been shown to date to around 800 BCE, and indeed it is the beginning of construction of these mounds that archaeologists have used to mark the beginning of Adena culture (Just as the building of New Cahokia marks the beginning of Mississippian culture). The Adena culture seems to have existed in Ohio for a long period, from its beginnings in 800 BCE until 1200 years later in 400 CE.

Seemingly the Adena had a similar fate to the people of New Cahokia in the early 15th century. The abandonment of this once great city of Cahokia was abrupt, dramatic, and complete. One day there was a thriving and robust metropolis, and seemingly the next there was a vast ghost town. What happened to Cahokia? This is a question that has exercised archaeologists since they first began work on the site, and there have been a number of different views, ranging from internal warfare to droughts and other means of food shortage. Maybe the simplest explanation is also one that now seems most likely, given new evidence: flood. The fatal flaw of many of the Mississippian cities and towns that grew up alongside of major rivers is that, while they were perfectly placed to take advantage of the superb agricultural benefits offered by placement in the floodplain of a river, they were also perfectly placed to feel the full brunt of the destructive force of that river on those times it overflowed its banks and flooded the surrounding countryside. Cahokia is located in a perfect spot to be decimated by a particularly bad flood—the kind of "hundred-year flood" we hear about occasionally taking place along major rivers like the Mississippi, and which wreak havoc on riverside communities. There is new evidence that just such a flood took place along the Mississippi River in this region during the period Cahokia was abandoned.[37] What started with a bang (SN 1054), ended with a fizzle.

There was a similar abandonment at the SunWatch site in Dayton SunWatch as it stands today is even *more* exposed to flood danger than Cahokia, as it sits 200 yards from the Great Miami River, which is particularly flood-prone in the Dayton area, where it intersects with the Mad River. In early spring of 1913, the city of Dayton was devastated in one of the worst floods in American history, which left all of downtown and the surrounding areas completely underwater. The area occupied by SunWatch was completely submerged. It may have been just such a flood that

[37]Munoz, et al. 2014.

led to the abandonment of the SunWatch site (though there is currently no evidence for this, and there are many other possibilities for its abandonment as well). Indeed, the natives of this region, with the benefit of hindsight, did not have settlements in what is now the Dayton area when Euro-Americans moved into the region to establish homes. A number of natives apparently warned early Euro-American settlers in the Dayton area that it was not a good place for settlement, because it was flood-prone. Dayton learned this the hard way in 1913. Today, there are tall levees alongside the Great Miami across its span through the city.

Is it possible that the Mississippian culture that developed on the banks of the Mississippi River, and had expressions in branch cultures throughout the eastern half of the current United States, was influenced by Mesoamerican culture, as some archaeologists believe? This is certainly far from impossible. If we are looking for the most likely sources of influence, there could me many. Trade, of course, is always an incentive to travel, and trade happens between cultures who otherwise never contact one another. Mexica artisans in Teotihuacan trade with groups further north, which themselves trade with Puebloans, and the artifacts of the Mexico valley end up in Chaco Canyon (current day northern New Mexico), for example. This is a standard story about influence. There is also more direct influence. The Maya city of Chichen Itza, for example, spent the greatest part of its existence (and its highest achievement as cultural center) not as a Maya-controlled city, but as a Toltec-Maya city, controlled by Toltec invaders from the western city of Tula in the Mexico valley. The Toltecs arrived in the northern Yucatan peninsula not by the (longer) land route, but by sea, sailing directly across the Gulf of Mexico that separates the Yucatan from the mainland. There are images at Chichen Itza and elsewhere depicting the Toltec in boats, surveilling the Yucatan shore, which they attacked from the sea. The Toltec people were clearly adept sailors. It is not a huge leap to think that perhaps such a seafaring people may have also sailed north on the Gulf of Mexico and landed along the Gulf Coast of the modern day United States. While the "Toltec Mounds" in modern day Arkansas, which were originally thought by some to be constructed by the Toltecs (and thus took the name 'Toltec') are not in fact Toltec at all, it may be possible that there was such northern influence. And if not by sea, then perhaps by land.

If there was Mesoamerican influence in the Mississippian area, this would be one explanation for the centrality of astronomy in ritual culture in both areas, as well as the similarity of a number of other cultural features.

References

Aveni, Anthony, 2001, *Skywatchers* (Austin: University of Texas Press)

Colas, Pierre, 2003, "K'inich and King: Naming Self and Person Among Classic Maya Rulers", *Ancient Mesoamerica* 14, pp. 269–283

Cook, Robert and Lane Fargher, 2008, "The Incorporation of Mississippian Traditions into Fort Ancient Societies: A Preliminary View of the Shift to Shell-Tempered Pottery Use in the Middle Ohio Valley", *Southeastern Archaeology*

Diehl, Richard, 2004, *The Olmecs: America's First Civilization* (New York: Thames and Hudson)
Friedel, David, Linda Schele and Joy Parker, 1993, *Maya Cosmos: Three Thousand Years on the Shaman's Path* (New York: William Morrow and Co.)
Kelley, David and Eugene Milone, 2005, *Exploring Ancient Skies: An Encyclopedic Survey of Archaeoastronomy* (Dordrecht: Springer)
Looney, Jefferson J., ed, 2009, *The Papers of Thomas Jefferson*, Retirement Series, vol. 6, *11 March to 27 November 1813*, (Princeton: Princeton University Press) pp. 322–330
Lounsbury, Floyd, 1982, "Astronomical Knowledge and its Uses at Bonampak", in Anthony Aveni (ed.) *Archaeoastronomy in the New World* (Cambridge: Cambridge University Press); pp. 143–168
Malmstrom, Vincent, 1978, "Architecture, Astronomy, and Calendrics in Pre-Columbian Mesoamerica" in R.A. Williamson (ed.), *Archaeoastronomy in the Americas* (Los Altos: Ballena Press)
Milbrath, Susan, 1999, *Star Gods of the Maya* (Austin: University of Texas Press)
Munoz SE, Schroeder S, Fike DA, Williams JW, 2014, A record of sustained prehistoric and historic land use from the Cahokia region, Illinois, USA. Geology 42(6): 499–502
Pauketat, Timothy, 1997, "Cahokian Political Economy" in Pauketat and Emerson, eds. *Cahokia: Domination and Ideology in the Mississippian World* 30-51 (Lincoln: University of Nebraska Press)
Pauketat, Timothy, 2009, *Cahokia: Ancient America's Great City on the Mississippi* (New York: Penguin Books)
Saturno, W, Karl Taube and David Stuart, 2005, *The Murals of San Bartolo, Guatemala, Part I: The North Wall*, Ancient America, 7 (Barnardsville, NC: Center for Ancient American Studies)
Schele, Linda and David Friedel, 1990, *A Forest of Kings* (New York: HarperCollins)
Taube, Karl, 1992, *The Major Gods of Ancient Yucatan* (Washington D.C.: Dumbarton Oaks)
Tedlock, Dennis, 1996, *Popol Vuh*: *The Definitive Edition of the Mayan Book of the Dawn of Life and the Glories of Gods and Kings* (New York: Simon and Schuster)

Chapter 3
Supernova–Cataclysm in the Sky

Introduction

We can only imagine the spectacle, and perhaps the terror of the people of the 11th century who looked up into the skies one night to discover a new star, brighter even than Venus, rivaling the crescent moon beside which it stood motionless and fixed along the background of stars. Perhaps no one living had ever seen such a thing before, and it must have mystified and challenged astronomically inclined peoples such as the Mississippians, the Maya, and the Chinese, who devoted whole classes of people to following the changes of the heavens daily and nightly.

The supernova of 1054, as startling and inexplicable as it must have been to those who saw it first hand, including the Chinese observers who chronicled it and most likely the priests and citizens of the old city of Cahokia, the shamans of the region that would become Chaco Canyon, the celebrants at current day Serpent Mound in Ohio, and elsewhere in the Americas, is something we understand much better today. SN 1054 A is the designation we today give this event, and it was a Type II Supernova, the explosion (or rather colossal disintegration) of a red or blue giant star. It is an event rare to witness with the naked eye, and those in the northern hemisphere in this year were treated to a once-in-many-lifetimes event. There have been no such supernovae visible during the lifetime of anyone on Earth today, or even our great grandparents. The last such event to be recorded was the Supernova of Kepler in 1604. This was also the last event recorded in the Milky Way by any means available to us today, visible or otherwise. No one knows for sure when the next such event will take place, but there are a few candidate stars that astrophysicists predict will become supernovae at least in the next thousand years or so (the blink of an eye in astronomical time!), and could happen at any time between now and then. Perhaps we will see such stars in our lifetimes! One can only hope.

For northern hemisphere observers, the best shot is probably the star Betelgeuse, or Alpha Orionis (the star designated alpha in the Orion constellation), one of the

A. McLeod, *Astronomy in the Ancient World*,
Historical & Cultural Astronomy, DOI 10.1007/978-3-319-23600-1_3

prominent stars of the well-known constellation Orion, which dominates the late fall and winter skies. Betelgeuse appears today as a wildly unstable red supergiant star on the brink of disintegration into a supernova (spectral type M21ab, which I will explain below). For southern hemisphere observers, the star Eta Carinae is most likely to appear as the next supernova of this type. Most skywatchers believe that the southern hemisphere has the edge in this race, as Eta Carinae appears closer to supernova than does Betelgeuse. Eta Carinae could also appear as a supernova anytime between today and one thousand years from now.

Of course, when we say that these stars could become supernovae "overnight", we have to be clear about what we mean. Eta Carinae, for example, is 7500 light years from Earth. One thing this means is that the light emitted from the star takes 7500 years to reach our planet. Thus, if we were to see Eta Carinae suddenly appear as a supernova tonight, that would mean that the star went supernova 7500 years ago, well before the earliest human civilizations, and we are only now witnessing the delayed footage, if you will, of the star's violent end. Whenever Eta Carinae appears in southern skies as a bright supernova, one thing that is clear is that the star has *already* become a supernova, as we know that the images we have are from a star that has nowhere near 7500 years left.

The term 'nova' is borrowed directly from the Latin, for 'new', and was applied to such events we now call *novae* and *supernovae* (using the Latin rules of pluralization) most famously by the 16th century observational astronomer Tycho Brahe, who documented the "new star" of 1572 and thereby helped to shatter the Aristotelian view dominant throughout Europe of the "perfection of the heavens"—a view that entailed that no changes could happen in the stars. The stars and the planets, according to the Aristotelian view, are fixed to crystalline spheres, and their motions are mechanical and eternally revolving, in motion. As perfect, the spheres not only do not, but indeed *cannot* change. Whatever undergoes change is less than perfect, as change entails destruction and decay. Only non-eternal things can change, but by definition the heavens are eternal—there has been no beginning of the cosmos nor will there be an end—and so it follows that the heavens cannot change. There can be nothing new or aberrant in the skies we observe.

Of course, there are a number of ways one can attempt to make the observation of unusual and irregular phenomena such as supernovae consistent with this belief in the changelessness of the cosmos. One way, which some astronomers indeed endorsed, was to claim that such seemingly astronomical phenomena as novae, comets, and other irregularities, were actually not astronomical at all, but meteorological. Another possible way was to deny that such features like supernovae were actually *nova* (new) at all. Instead, one could claim that such instances were results of relatively infrequent, and thus less observed, alignment phenomena based on the existing spheres. So, for example, perhaps given the motions of the planetary spheres within the background sphere of the stars, every few hundred years the spheres line up such that the particular correspondence creates a star-like effect in the sky. Thus, what we saw in 1054 was not a supernova at all, but simply a regular (though long period) effect of the normal and mechanical motion of the spheres.

While this explanation would perhaps have been more intellectually satisfying, it conflicted with the Ptolemaic tables in the *Almagest*, which was the end-all authority for most of medieval and into modern European history.

Why, we might ask, would one think as did Aristotle, and Ptolemy after him, that only a changeless thing can be eternal, and that the cosmos must itself be such a thing? The answer requires a look to cherished ancient Greek ideas about change, dating back all the way to the Presocratic philosophers (we will take a look at part of this story in Chap. 8). Briefly, one might argue that any change is the coming to exist of one quality or thing and the ceasing to exist of another. If a thing can cease to exist, then it must cease to exist due to some condition or other—it does not just randomly cease to exist. But then this means that the existent thing that can go out of existence—that is, can change, is dependent on conditions, and anything that is dependent on conditions cannot be, by definition, eternal. The eternal is that which *always* exists, independently of conditions. Since things that change cannot do this, then changing things cannot be eternal. The cosmos itself is certainly something eternal, the ancient Greeks thought, because it exists independently of conditions. It is the basis for the existence of everything else, and there are no conditions under which the cosmos would cease to exist. And if it is eternal, it does not then change. This is why, according to Aristotle and his intellectual heirs, we see the regularities that we do in the cosmos. But of course there is a problem. Supernovae, comets, eclipses, and other apparently non-regular "changes" in the cosmos. Aristotle and later Aristotelians struggled mightily to explain away these phenomena, most often dismissing them as merely meteorological events. The sky might change, but the celestial spheres could not.

As it turns out, those alive during the late sixteenth and early seventeenth centuries, such as Johannes Kepler, were given a rare astronomical gift—two supernovae within a single lifetime. The supernova of 1604, also commonly referred to as "Kepler's star", after the astronomer's pioneering work on the supernova, followed a mere 32 years after Tycho Brahe's supernova of 1572. Brahe unfortunately was not himself alive to witness this second supernova, having died just a few years before, in 1601. Brahe had, however, conclusively demonstrated that the supernova of 1572, and presumably all other supernovae, including the later SN 1604, were indeed neither meteorological phenomena nor side-effects of the motions of the spheres.

Brahe did this by determining the distance of the new star, in relation to the moon and the other planets. While Brahe was unable to determine the exact distance to the supernova—indeed no one would be able to determine stellar distances in this way until the 19th century, and not with regularity until the 20th—he was able to determine that the object was further than the moon or the stars, and thus could not be a meteorological phenomenon. He demonstrated this using the technique known as triangulation, which attempts to detect the phenomenon of *parallax*.

The parallax effect can be described simply. If you hold out a finger in front of your face, about two feet or so, and focus on it, it appears in one place against the background environment. If you now close one eye and look at it, you will see it from a particular location against the background. But, keeping your finger in the same place, if you now open your closed eye and close the other eye, still looking at the finger, you will notice that it appears to have shifted its location against the background. It will appear to the right or the left (depending on which eye you closed) of the location you first observed it. This is the parallax effect. The cause of this effect is based on the fact that the two observations are made from different locations, and, while the observed object hasn't moved, the observer has moved, and thus views the object from different angles relative to the background environment. In the case of the finger-and-eye example, the two eyes can be taken as different observation points. I observe my finger from the vantage point of my left eye, then I observe it from the vantage point of my right eye, a few inches to the right (Fig. 3.1).

The parallax effect is increased based on two factors: (1) how close the object being observed is, and (2) how much distance perpendicular to the line bisecting the two observation points there is. Thus, if you view an object against the background standing 50 feet in front of you from two positions a foot apart, you will notice a considerable parallax shift. (Indeed, even if you try viewing such an object from one eye then the other as in the finger example above, you will notice a slight, but present, parallax shift. As I write I am looking out my window at a tree about 50 feet in the distance on a cold fall morning, and observing it through different eyes I find a parallax shift about the width of a leaf 50 feet further in the distance. The leaf is visible with my left eye open, its edges touching the bark of the tree, and invisible with my right eye open, with the leaf next to it still visible.) If you now try to view an object 500 feet away against its background, you will not notice any parallax shift when you observe the object from one eye then the other—although there *is* a parallax shift even here, just one that is far too small for our visual system to detect. There will be a parallax shift for *any* observation of any object against a background

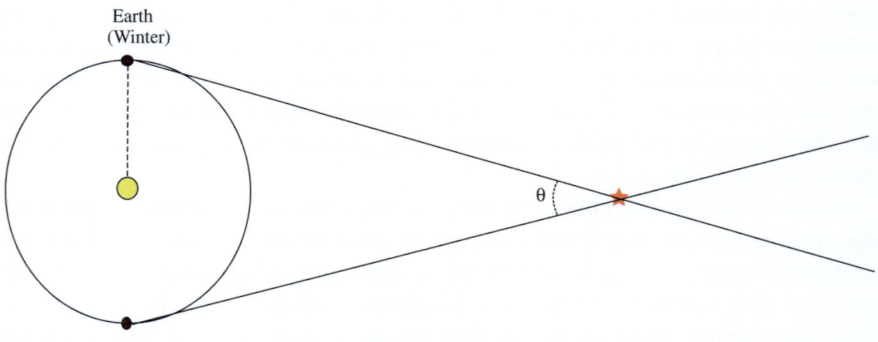

Fig. 3.1 The Crab Nebula (M1) is the remnant of the enormous supernova of 1054. Its Messier number signals the fact that it was the first deep-space object catalogued by 18th century astronomer Charles Messier. Messier's original list of objects, anchored by the Crab Nebula, contained 45 objects. It grew in later years to 109 objects, 103 of which were catalogued by Messier himself. The Messier objects are popular targets for amateur astronomers, as they can all be resolved even with relatively small amateur telescopes from dark sky sites (Photo credit: NASA (Hubble Space Telecsope mosaic))

from two points not along a line of sight to the object. The issue is whether we can *detect* such a shift. Some will be so small that we cannot detect them, such as the parallax shift of Alpha Centauri against the background viewed from one eye then the other.

The method of triangulation used to detect parallax can be used on celestial objects as well. For relatively nearby objects such as the moon, the two observation points can be on different parts of the Earth, and this will suffice. For more distant objects, this will not work.

For objects such as the planets, the situation is trickier. It is possible to observe an object like a planet from two positions, as the Earth turns in its orbit around the sun. Observing at January 1st, then again at March 1st, will give two different positions, as the Earth will have moved 1/6th (more or less, given that the Earth's orbit is elliptical, and the planet covers more space sometimes than others) of the way of its orbit around the sun, or roughly 97 million miles. That's a good enough distance to measure parallax on many celestial objects. With planets, we have the additional complication that they are moving in their own orbits, and so we have to know how much distance is covered in those orbits relative to Earth motion in order to determine parallax shift. But this can be done, and could be done long before we were able to detect parallax shift in anything as distant as a star.

For many years, the fact that we could not detect parallax shift in any of the stars was taken as support for the Aristotelian/Ptolemaic theory of the fixed stars as an outer sphere of unmoving objects. They *were* the background, within which everything in the universe, our solar system, happened. There were no changes in the realm of the stars, but they were eternal, fixed, and changeless. Tycho Brahe demonstrated conclusively the falsity of this view when he determined that the "new star" of 1572, like the other fixed stars, was indeed not a meteorological event, or even something within our solar system. It had no detectable parallax shift. The perfect and eternal fixed sphere of the stars had *changed*. The Aristotelian notion of the perfection of the spheres could not be maintained.

What neither Tycho Brahe nor his contemporaries who accepted the Ptolemaic view could determine, of course, was the cause of this new star, which was just as much a mystery to them as the new star of 1054 had likely been to the Mississippian astronomers of Cahokia and the Mayan astronomers of Chichen Itza.

A supernova, according to the contemporary scientific understanding, is the massive and violent explosion of a star of sufficient mass, distinct from the smaller phenomenon of the *nova*. Although we call a supernova a single thing, there are actually different types of supernova, that are very different processes. One is an event more like the result of disintegration than a pressure explosion, in which the integrity of the star is no longer sufficient to contain the nuclear process happening in the center of the star, and the ejecta from the process launch into space rather than being contained and recycled by the mass of the star.

The other way is the more typical way for a celestial explosion to take place. The core of a massive star can no longer be supported, and it implodes inward, with the increased mass resulting in a nuclear reaction that cannot be contained by the mass of the existing star, and there is an explosion. In order to result in a supernova of this type, a star must be sufficiently massive, at least about eight times larger than our own sun, which is fated not to end life as a supernova, but instead burgeon out into a red giant, swallowing the earth before scattering into a planetary nebula, of the kind we can see on a dark, clear night through a telescope, such as the Ring Nebula (M 57) in the Lyra constellation in the late spring and summer sky, or the Dumbbell Nebula (M 27)[1] in the Vulpecula constellation.

Astronomers distinguish between two main types of supernova, Type I and Type II, and there are sub-types within these categories. The distinction between the two types has to do with the observed absorption lines in the spectra of stars, rather than the cause of the supernova (although it turns out there is generally a correlation between causes and spectral types). The standard distinction is that the spectra of Type I supernovae do not contain hydrogen in their absorption lines, and the spectra of Types II do contain hydrogen. So what does this amount to, concerning causes? Well, for all supernovae except for Type Ia, the ultimate cause is a collapse of the core of the star, issuing in explosion. Type Ia supernovae are an interesting oddity. A white dwarf, by itself, is not nearly massive enough to ever issue in a supernova. One day in the distant future, long after our sun has sent us all to *xibalba* and engulfed the solar system, it will dissipate into a planetary nebula. This nebula will slowly condense into a white dwarf star—a small, dense star in which there is no nuclear fusion, as its fuel has been exhausted. When you look up into the night sky with a good enough telescope at the Ring Nebula, you can see a tiny white dot at the center of the inner ring. This is the white dwarf that is all that remains of the once sunlike star that became the Ring Nebula. Perhaps there was once an earthlike planet around this star, itself with creatures who built temples to observe the skies, who struggled to understand their place in the universe, and understood the cosmos as central to their own lives.

The size of a white dwarf is dramatically smaller than that of the sun. In fact, its size is about the same as that of the Earth. At the same time, its mass is roughly equivalent to that of the sun, which makes it an incredibly dense object. We might imagine a white dwarf as a star of the sun's mass crunched down into an area the size of the Earth. While these stars are very small and dense, they can also be extremely bright. In fact, one of the best known stars, as well as the brightest star, in our night sky, is a white dwarf. Sirius, or the "Dog Star", in the Canis Major constellation, is a historically important star, and also one of the closest stars in our neighborhood of the Milky Way. It also, it turns out (and no one knew this until the modern age) is actually not *one* star, but *two*. Sirius is actually a binary system, comprised of two stars called (unimaginatively, because this was only discovered in the scientific age) Sirius A and Sirius B. It is just in such binary systems (where at

[1]Deep sky objects such as nebulae, galaxies, and star clusters, at least the most well known and visible ones, often have an associated *Messier* number (such as M 57, M 27, etc.). This is after the index of French astronomer Charles Messier (1740–1817), who indexed these objects in his catalogues, the final of which was published in 1781, and included 103 objects. Today, people refer to *110* Messier objects, as after Messier's death scholars discovered that he may have seen and catalogued seven additional objects.

least one member is a white dwarf) that the Type Ia supernova is possible. In some binary systems, a white dwarf begins to pull material from a disintegrating aging star, in the process of expanding outward and dissipating into a nebula, and when enough material is gained by the white dwarf, the process of nuclear fusion that ended in the star that is now a white dwarf will be restarted, with catastrophic effect, the supernova explosion. These supernovae tend to be very bright, like white dwarves themselves. And it was just this type of supernova that was observed by Tycho Brahe in 1574 and helped to demonstrate the falsity of the Aristotelian view of the heavens, and by Johannes Kepler a mere 30 years later in 1604.

Our supernova—the supernova of Cahokia and the Postclassic Maya, the supernova which we will see may have had important historical effects, was a Type II supernova. The Chinese astronomers, who we will look at much more closely in the next section, were, as they often were with many other astronomical events, the first to record observation of the star. The Chinese account mentions that the star was visible during the day, clearly brighter than Venus, and was visible for 23 days.

The remnants of the supernova of 1054 can still be seen today. According to the western constellations, near the lower part of the constellation of Taurus (the star Zeta Tauri), one finds the object known as the Crab Nebula, or M1 (the first catalogued object in Messier's list). According to Messier, it was discovery of this object that led him to compile his catalogue, so as to help astronomers distinguish between these strange (today known as "deep sky") objects and closer and more variable objects like comets. The nebula itself is an irregularly shaped splotch in the sky. It *looks* like the remnants of a violent explosion, unlike the more often smoothly rounded planetary nebulae that result from the slow expansion and dissipation of sunlike stars. One can see it on a dark and clear night when Taurus is visible, with the aid of a telescope of at least about 6 in. aperture. While the magnitude of the Crab Nebula as it appears today is about 9,[2] in 1054 when it appeared as a supernova, the magnitude was likely somewhere around −5 or −6, brighter than Venus. This illustrates another fact about SN 1054, now the Crab Nebula—it is slowly fading. The Crab Nebula is the visible remnant of that supernova, and over time it will dissipate, condense, and eventually continue on in its evolutionary cycle. Thus, the Crab Nebula itself is dimming over time. In the days just after SN 1054 became invisible to the naked eye, it would have appeared clearly for those with small telescopes as a bright deep sky object, perhaps magnitude 6 or so. Today, it has dimmed to magnitude 9. Eventually it will diminish in magnitude to the point that it can only be detected by the most advanced telescopes and will no longer be resolvable by amateurs. Fortunately this is far off in the future,

---

[2]The lower the magnitude, the higher the brightness. Thus, Venus has a variable magnitude of about −4.9 to −3.8, depending on its position relative to us, and the brightest stars such as Sirius has a magnitude of −1.46. In a completely clear and dark sky, the limit of magnitude visible by the naked eye is probably around magnitude 4 or 5 (for those with Eagle eyes), and in light polluted areas this decreases considerably. Most nights in my own home, one is lucky to see stars around magnitude 2, even on the clearest of nights. The Crab Nebula is far too faint to see even from a good telescope in my own backyard.

Fig. 3.2 Parallax. (Illustration by the author)

well past the lifetimes of any of us, our children, or our lineage down to posterity. And in the long run the sun will engulf the earth, after all, so we probably have plenty of more pressing issues to worry about (Fig. 3.2).

SN 1054 Elsewhere in the World

SN 1054 is controversial, in its apparent historical connections, mainly because we do not know much about it, and cannot definitively make the link to a number of events of import around this time, including the shift of Cahokian life and society, the cave paintings in Chaco Canyon and elsewhere, and a massive shift in European life. In 1054, after years of distrust and sometimes outright hostility over differences ranging from fine points of theology to more mundane and secular issues such as authority and who has power over what, the Eastern and Western branches of the heretofore unified orthodox Christian church split from one another in what came to be called the "Great Schism" (or East–West Schism, so as to distinguish it from the distinct schism that happened in the West with the Protestant Reformation). This schism is maintained down to the current day, and is the reason that the Roman Catholic Church and the Eastern Orthodox Church are distinct. They were once, prior to 1054, the same Church. While it was true to say that there had still existed these two factions, or branches of the Church before 1054, one comprised of those under the authority of the Patriarch of Constantinople, and one of those under the authority of the Pope of Rome (notice the continuity between this distinction and that between the eastern and western branches of the Roman Empire—and the similarities do not stop there—Latin was the official language of the Western Church/ empire, Greek of the Eastern, etc.), they were at least in letter of the law part of the same Church. 1054 dissolved this tenuous union.

Was the 1054 supernova instrumental in bringing about this final break between East and West? Interestingly, and perhaps bafflingly, there is not a single existing account of the supernova in the European records. It is strange to think that *no one* wrote about the supernova that everyone must have clearly seen, visible even during the daytime. This leads one to think that they must have either been avoiding noting it for a specific reason, or that the records have been lost. If the former is true, this could have to do with the Schism, as officials wanted to avoid documenting anything that made the Schism seem divinely inspired, or to give fodder to the other side. Of course, if the supernova did help to *occasion* the split between the churches, one would expect that each side would offer accounts of the celestial event, complete with interpretations supporting their own side and maligning the other—claiming that the star, like the star of Bethlehem that appeared at the birth of Christ (possibly another historical supernova), signaled a new birth, the freedom or realization of the true Church, and the purge of pretenders—or some such tale.

No such account exists. The only certain account of the supernova that exists today is that of the Chinese court astronomers—and it is still not completely certain to some that the 1054 supernova that they recorded can be identified with the Crab Nebula. We are fairly sure they are the same, but there is some discrepancy in the account. The Chinese accounts claim the 1054 supernova to occur in a slightly different portion of the sky than we find the Crab Nebula today. For those knowledgeable about Chinese astronomy, it is relatively easy to explain why this might be. One of the primary tasks of astronomers in early China was to read the stars and give portents (an activity we might call 'astrology'). Imperial astronomers were responsible for determining the fate of the dynasty, by reading it in the stars. There would have been more than a little pressure on astronomers to make sure that enormously significant celestial events like the 1054 supernova were read not as auguries of doom for the dynasty, but as signs of the favor of Heaven. This was a matter of the highest delicacy and import, as any sign that a celestial event could be read as an indication of Heaven's displeasure could embolden rebels and spark a chaos that could very possibly result (as it sometimes did) in the end of a dynasty. Thus, Chinese astronomers might be tempted to slightly change the positions of certain events in their records, so that they don't impinge on some important star associated with imperial favor or success. There is some indication that this was the case with the 1054 supernova.

While there is this single solid account of the 1054 supernova, the Chaco Canyon inscriptions of the Anasazi and others in the region may constitute a second existing account of the supernova. We cannot establish this with any degree of confidence, however, as it may be the case that these inscriptions depict Venus, or a common image of a star and crescent of the type we see around the world (consider the symbolic imagery of the star and crescent in the Islamic religion, for example).

The supernova of 1054 was hardly the first or the only supernova to make its presence felt in human history. There were other much better documented supernovae to make their entrance in human history, and have an enormous effect on it. Particularly well known are the supernovae of 1572 and 1604 mentioned above, which helped inaugurate the age of contemporary science in the west, and which doomed the views of Aristotle and Ptolemy that had been accepted almost as holy writ in Europe for so long.

While our supernova, that of 1054, was a Type II, there had been another supernova visible from Earth less than 50 years earlier, in 1006. Thus Type Ia supernova was much better documented than that of 1054, and for a good reason: this was likely the brightest non-solar celestial phenomenon ever witnessed in human history. Unlike with SN 1054, it is not only the Chinese astronomers who documented this supernova (although the Chinese were the masters of observation in the premodern world—generally if they didn't document it, it either wasn't visible to them, or it didn't happen), but Arabic astronomers also documented it, and likely also Hohokam peoples in what is today the southwestern United States. During this period, there is very little recorded astronomical observation from Europe, which was smack in the middle of its "Dark Ages" (which is, of course, as any historian will point out, somewhat unfair. No culture just goes dark, but different things are prioritized, and in Europe religion, theology and political dispute were more important than astronomy, which was neglected. Of course, we can justly criticize any culture as well, and by my lights lacking concern for astronomy qualifies European cultures at this point for the moniker). For some reason, likely a coincidence (but who knows?), supernovae visible to us seem to come in pairs, relatively close together in time, followed by a long absent period. For example, there were supernovae in 386 and 393, then none until 1006 and 1054, and another long hiatus until the supernovae of 1574 and 1604.

Today, supernovae can be observed with regularity, but only because of the aid of modern technology. Before the creation of telescopes not only that allow us to resolve distant objects, but that also can detect light in different wavelengths, including infrared, ultraviolet, as well as telescopes detect radio waves, we were only able to observe "naked eye" supernovae that can be seen readily by anyone on the ground in the right part of the Earth. We have not witnessed one of these events since Kepler's supernova of 1604. We find so many new supernovae using modern technology primarily because these supernovae are *extragalactic*—that is, they happen in galaxies outside our own. Before the first supernova discovered by telescope was seen, all of the visible supernovae had been within our own galaxy, and we had never witnessed an extragalactic supernova. These events are so bright that when one happens in our own galaxy (for the most part, there are some exceptions) we will generally be able to see it without the aid of a telescope. At the same time, even the brightest extragalactic supernovae will require sophisticated equipment to observe, given its enormous distance and relative small size.

Generalizing from the data, supernovae in our own galaxy take place about once or twice every century (though of those a smaller number will be visible), and this explains the low incidence of observation of supernovae unaided by telescope in human history. When we add to this all of the other galaxies that become available with the aid of contemporary telescopes, which are improving all the time, the number of supernovae observations rises exponentially. The latest significant supernova accessible to even some amateurs happened in 2011, in the Pinwheel Galaxy (M101), 21 million light years away. A type Ia supernova appeared in one of the arms of the galaxy, reaching magnitude 11, bright enough to be seen in a decent amateur telescope from a dark sky site. However in most developed areas, one cannot hope to resolve anything approaching magnitude 11. The limiting magnitude in my own observation, even in relatively dark sky sites in various places in the U.S., has been magnitude 7 or 8. Of course, the more distant the stars we observe, in remote galaxies, the further back in time we are looking. The light that reaches us from the most distant galaxies we can observe with the most sophisticated tools began its journey from its home galaxy about 13 billion years ago. This is an amazing number, when we realize that our universe itself is only 13.8 billion years old. To view such distant galaxies, then, is to look back in time to the infancy of our universe.

The Philosophy of Time—On Time, the Nature of Time, and Its Changing

Which brings us back to the issue of *time*, and its connection to the sky. In the last chapter we looked at the famous Long Count calendar of the Maya. In the Preclassic Period, the Long Count was not in use on urban inscriptions as it came to be during the Classic Period, and the rise of this use of the Long Count, as well as reforms in language, came with the institution of the concept of divine kingship in the Classic Period. During this period we see the elevation of rulers to identification with the sun, as discussed above, and the rise of the view that the ruler, the *ahaw* (lord) is not only lord of his land and people, but lord of time and space. It is for this reason that many of the urban inscriptions and even the images from the few available surviving texts depict kings and their various exploits, especially conquests, feats of power, or divine representations.

This cannot be all there was to the story of the Maya renaissance, of course, but the ruler clearly had enough power to employ the entire creative class in works focused on him and his power and glory—similar to what we see in many societies in Eurasia at the time.

Time is a complex issue in any culture, and the philosophical and religious roots of any culture can often be found in its conception of time. The Maya conception of time is one that has echoes in certain ancient Indian, Central Asian, and African views, but one that is foreign to many other views of time, including dominant views in Europe and East Asia. According to the Maya conception, each period of time contains within it both the echoes of the previous period and the traces or anticipation of the future period. Moments are not discrete and separable. The present does not exist alone on its own, ontologically distinct from the past and future. Views that hold the present to be separable, such as that of Aristotle, were popular in the West, and amount to the position that only the present exists. There is only now—the past no longer exists, and the future does not yet exist. Perhaps we can locate the modern injunction to "focus on the present" or for the more active-minded "seize the day" as arising from such a conception of time. The Maya conception contrasts with this kind of presentism. The past and the future both not only exist, but have important effect and support in the present. The reason I do what I do is based in the past, which is acted out now in the present, and the future also affects my actions, when I do what I do with an eye to it. The future does not exist off in some nonexistent void, but before my eyes. If it did not, how could I do what I do in part because of it?

We can see that a major concern of Native American astronomy in general is with keeping time, with dictating calendars, in terms of not just agricultural periods, but also ritual occasions, political and social organization, and historical marking, among other things. We are lucky enough to have particular accounts of the calendars of one such civilization in the case of the Maya, and studying the Maya conception of time and its basis in the sky can tell us both what ancient civilizations in the Americas thought about the sky, as well as how they thought about time. Today, we tend to take the issue of time for granted, thinking that it is a single thing inexorably moving forward, a progression from present to future, and present into past, dictated by the ticking of the clock, and the flipping of days in the Gregorian calendar. But time has never been such a simple concept, and is no less complex today than it was for the Maya in the Classic Period.

But what is time, and how do we make sense of its passing? Certain views and assumptions in the philosophy of time will make a great deal of difference to how a culture approaches both the calendar and the sky. Time is something that all of us experience, or appear to experience, and have and intimate association with and knowledge of—we think. Time passes and exists the same for everyone, and there is a universal standard, thus making the reality about time independent from culture, individual experience, or other subjective features. The problem with this picture of time is that it cannot be true. How do we associate time with all of these above things, and how can it be that time is linked with some objective and external standard?

Consider the passing of time. How do we *measure* this objective time? If time is independent of all experience or appearance, and then independent of measure for that matter, what is it that we are measuring when we measure time? With the tick of a clock, or the view of one rotation of the sun in the sky, is it something independent from these that is being measured? What could it possibly be? Consider the most accurate clock, ticking its way through the day. With each of these clicks, what is being measured? Think about how we might know that the watch is off or inaccurate? It keeps time well at first, and then becomes slow. How do we determine this? Perhaps it loses time with other watches, or with the motions of the sun, or with the atomic clock. But notice that what we compare it to in order to discover whether it is accurately keeping time are *other* time pieces, not time itself. But if the watch is supposed to be a measure of time itself, then what are we doing here? If time is an objective and independent thing, we should be able to measure time directly, right? Well, that depends on what *kind* of thing time is. And just how independent and objective it is will also depend on this.

Notice that whatever we choose as a time piece, there will always be an independent question as to its accuracy, and to determine its accuracy our only option will be to measure our time piece against other timepieces. Say that we use the motions of the sun in the sky as a measure of time. How do we know that we are consistently keeping track of days—that is, how do we know that one day isn't slower or faster than another, which we could not know by consulting our timepiece itself? We can only defer to other more trusted timepieces. Maybe we have access to a water clock, which we think is regular in its operation, or an atomic clock, both of which show us that the sun's rounds actually do not give us consistent days—some are longer than others. Of course, we can only determine that these days are unequal in length *in comparison* with the atomic clock or the water clock. We have no independent way to determine the accuracy of the atomic clock against time itself. If the earth's rotation is the most regular action in the universe, then any discrepancy between it and the atomic clock would mean that it is the *atomic clock* that is inaccurate anytime there is a discrepancy, not the opposite. The reason we take the opposite to be the case is that we take the process of atomic breakdown that constitutes the atomic clock to be a more regular process than that of the earth's rotation, which we have many reasons to believe is not very regular (comparatively), including the gravitational effects of objects such as the moon on the earth's rotation.

Ultimately, whether one time piece or the other is accurate is only determinable based on what we take to be more accurate timepieces, and this can only be determined based on our conceptions of which processes are more regular than others. There is certainly an enormous amount of subjectivity in this. Who is to say that the atomic process is more regular than the rotation of the earth (gravitational effects aside)? Maybe the gravity of the moon is slowing the atomic process! Of course, this all sounds absurd, but part of the point is that there is no way we could know this from the standpoint of time alone. If we know that the atomic clock is more regular than the rotation of the earth, it has to be due to facts that we discern about the irregularity of the process of the earth's rotation independently of consideration of time. The cultures of Mesoamerica and North America discussed in Chap. 1 had different conceptions of time in part because they adopted different standpoints from those we accept in the contemporary world. Astronomical phenomena, for the Maya, dictated a great deal of the content of the foundation on which they understood the basis and movement of time.

To further illustrate the problem with thinking about objective measures of time, imagine that all of the things we experience in time were sped up by many times. So our waking up and eating breakfast and walking out the door, instead of taking 20 min, would now take 10 s. Everything would moving fast-forward as in a sped up movie. How would we determine that we were moving faster? If we were watching a sped up movie, we would know that the movie was moving faster, because we would be able to compare it to the motions off the screen. Of course, notice that we would not really be able to determine whether the movie was moving faster, or whether we slowed down the entire rest of the world with our pressing of the fast-forward button, just based on a consideration of rate of motion. But what if *everything* were sped up like the movie? Now, you would not be able to determine the difference between the 20 min breakfast and the 2 s breakfast. Because not only does your meal pass that quickly, but also cars drive by your window faster, you read and understand the paper in an instant, the sun moves through the sky faster, etc. *Everything* has sped up by the same rate. One would be unable to tell the difference between things progressing at the normal rate and a sudden shift to the entire world moving 20 times faster. Your heart rate, cell movements, and even neuron firings have all increased in speed by twenty times. The reason we would be unable to tell the difference is because all of these things would still happen in the same time relative to one another. That is, one heartbeat would still take about the same time as one click on my watch, etc. So what should we say about such a case? If we would not know the difference, should we say that there has been no change? Or rather that we simply could not determine the change? After all, the two are not the same. We may not know what the interior of a black hole looks like, or what a particular star outside of our range of visible light is composed of, and may *never* be able to know these things—but we should not necessarily take this to show that there is no fact of the matter about them.

In the case of time, things are complicated. If what is being measured by time is nothing independent of the relative motions of things in the world, than it indeed doesn't make sense to say that *everything* has been sped up by 20 times. Things can only be sped up relative to other things. And those things slowed down relative to it. Time, that would mean, depends on frames of reference. And indeed this is just the claim we find justified by the theories of relativity, that take as core the constancy of the speed of light. The speed of light cannot be sped up or slowed down relative to anything else, its speed is constant and cannot be altered in any way, and so all other motion must change relative to it. There can be nothing we measure intrinsically with clocks or any other measure of time other than rates of change, as without this, there is no fact of the matter as to time. This is not simply a fact about our observability of time. If there were no change, there would just be no time, as time is just a way of keeping track of change. In a state of no change—no motion of atoms, nothing whatsoever, there would be no difference between 3 s and three millennia. Time is rate of decay, rate of motion, rate of change.

If time is nothing more than rate of change, then our selection of time-pieces says quite a bit about what processes, changes, and events we deem important. When one looks at our modern world, one can determine that we take regularity and precision as of importance, on the societal or institutional level, while on the level of smaller communities we have different ways of determining time. Hardly everyone consults the atomic clock to determine how they structure their day or year. We may use the seasons to determine time—residents of a beach town may think of things in terms of summer versus non-summer, to structure their lives around the period of economic activity in their town. A university student or professor will, as discussed above, structure their life around academic semesters or quarters.

We can ask the question: how did the ancient cultures discussed in this chapter structure their lives? How did they understand time? And we can give at least a partial answer, on the basis of the phenomena they used to determine time, which tells us something important about their concerns. The Maya, as we have seen, used both the sun and Venus, the hero-twins Hun Ahaw and Yax Balam, to determine time. It is fitting that these two time pieces represent the heroes, whose actions were responsible for the generation of humanity, according to the Maya. The sun's determination of time is clear. One day is a basic unit of time in almost every culture. The sun determines days, as well as seasons, which are also recognized in every culture I know of. The sun determines the times at which we can see outside, and the times at which we cannot, and so determines our times of sleep and waking. This is a primary and fundamental source of cyclical change at the center of the life of any individual or society, and so the sun becomes the most natural determinant of time. For the Maya it was no different, nor was it for the other cultures of the Americas, who invested much of their astronomical time and energy in observation of the various positions of the sun throughout the year.

The motions of Venus, the moon, and the sun did not, for the Maya, happen *in* time, rather they *defined* time. The subtle and complex interplay of these objects on their round through the sky constituted a changing structure around which one could organize one's life. It is likely this interplay that led to the interlocking physical calendars of the Maya, Aztecs, and cultures such as the Olmec before them. Time was not a matter of increasing intervals based on a *single* motion, such as that of the sun, or that of the atomic decay. Rather, these fundamental and important astronomical movements *together* determined a place in the calendar. For us, it may be significant that these movements of the celestial calendar overlap, and we see a system that appears to us strange. But the harmonies of the Mesoamerican calendars are apparent, and perhaps none more so than one feature that illustrates the genius of the Mesoamerican astronomers—the harmonization of seemingly disparate motions in the calendar. The Mesoamerican calendars not only integrate the sun and the moon's motions, as does our own Gregorian calendar, but also perhaps that of Venus and Jupiter. And they do so in a way that allows us to appreciate the amazing symmetry and complex beauty of the celestial motions, intuitively seeing the mathematics through which we can understand these motions.

Another central feature of the Mesoamerican calendars, and of Mesoamerican cultures in general, is the focus on the *cyclical* nature of time. Time should not be thought of as beginning in some point and moving through a line to end in some point, any more than the motion of the celestial bodies should be thought of in this way. Like these heavenly motions, time is a matter of cycles, and we can mark cycles through things such as alignments. As many experts on Maya history and culture have pointed out, the idea of an ending of time in 2012 at the completion of the last *baktun*, for those who understand ancient Maya culture, is absurd. The

Maya had a truly expansive notion of time, even broader than our own. There are numbers written in Maya texts that express the position that express the motion of time long before what we currently think is the date of the beginning of the universe, and long after what we believe will be its demise. This conception of time makes sense on the Maya cyclical conception. There was no zero point, but rather a series of changes. There may have been a zero *alignment*, but even insofar as there was anything like this, it was arbitrarily determined by humans. There is some evidence that the date August 13, 3114 BCE, which is the beginning date for the Long Count calendar, signifies just such an alignment. At the latitude of the southern Maya area, as well as much of the Oaxaca valley, where the calendar may have developed, one of the two zenith passages of the sun takes place on August 13th. It may then be that the "first" day in the expansive Long Count calendar was based on the positioning of the sun directly overhead in the zenith—but certainly the sun did not *begin* overhead—it rose that morning, and achieved this pinnacle of the sky. And, as we see in the Maya texts, the cycles determining time carried on long before this, and will carry on long after the end of the last *baktun* on December 21, 2012. Time will not stop any more than the motion of the celestial bodies will stop.

Our own conception of the operation of the heavens gives it a much more tenuous nature. In general, the connection between humanity and nature in the modern world is a much more problematic issue. While the Maya and the Cahokians understood that their heavens would continue to rotate, and the world that they always knew would continue to exist, long after their own deaths, in the modern world we face the stark reality that our own actions threaten the very existence of this world.

And why has this situation come about? We might argue that the very failure to see the rest of nature as continuous with and indeed part of humanity and the human realm has led to this crisis that we face today. It is of course not unique to the modern world to ecologically destroy oneself. There is evidence that such is what happened to the people of Easter Island, for example, and even perhaps the Classic Period Maya, or the people of Cahokia. Humans have an unfortunate tendency to display a complete lack of foresight, and even to do things that we know will likely lead to our future destruction, in order to attain present gains. We seem to have a serious inability to take the long view seriously, and this sometimes leads to complete disaster. Overfarming of fields leads to drought and starvation, overpopulation leads to disease and conflict. And we do not learn our lesson, even after seeing the same thing happen time and time again throughout history. But the situation can be exacerbated by particular philosophical views that excuse or even sanction the destructive habits of humanity. Our ethical philosophy should ideally be something that restrains, challenges, and reforms us, not something that simply offers a justification or support for how we are, accepting our natural faults. If our ethical philosophy does not ultimately help us to become *better*, than what use is it as a tool? (making a distinction here between ethical philosophy and other kinds—not *every* kind of philosophy has its primary value as a tool or something with some kind of instrumental value).

We should be careful, of course, not to read our own concerns into the thought of the Maya, Aztecs, Mississippians, and other Native American cultures. This is always a danger faced by contemporary students of these cultures. Because there is so much that needs to be pieced together, we often add our own prejudices or ideals, depending on our stance toward the culture in question. In the past, many people saw Native Americans as representing the "savage" and "primitive", and this conception was one that arose from a desire to see their own culture as the modern, scientific, and progressive future of civilization. Indeed, the very term 'civilization' is frought with this difficulty. On the other hand, there were (and are) some people who idealized the cultures of precolumbian America, imagining the native peoples of the Americas as peaceful, spiritual and philosophical reflectives, at one with nature, their world and the cosmos. The "noble savage" view that arose in the Americas with the colonizers is, perhaps, better than the "simply savage" view, but not much. Both miss the mark. And in any view, there is a tendency to read into the culture of those who we investigate the ideas that motivate us. So the environmentalist sees the natives of the Americas as people who revere and respect nature and the non-human, and ignore or explain away the fact that natives sometimes overfarmed themselves into civilizational collapse, overpopulated to unsustainability in their cites, transformed "pristine" areas through urban deforestation and "slash-and-burn" agriculture, and all kinds of other things the environmentalist condemns modern society for doing. The pacifist will sometimes see the native peoples as peaceful and non-warlike reflectives. Indeed, one early view of the Maya people, before we discovered better from deciphering texts and inscriptions, held just this about the Maya. Such views conveniently ignore the fact that warfare and violence was rife in the precolumbian Americas, every bit as much as it was in the 'Old World'. Maya cities, the sites of perhaps the most impressive cultural achievements of the Americas, were also the sites of bloody human sacrifice, including the sacrifice of innocent children (and it is absolutely not true that, as is sometimes claimed, those sacrificed were often willing or saw it as an honor—this is why captives from battle were most often used as sacrifices), and almost constant warfare, in the aid of the establishment of sometimes tyrannical, but always totalitarian kings. The stories of the Americas are as familiar as the stories of Eurasia—they are human stories. While there are cultural and philosophical differences between peoples and traditions, the basic patterns of human thought and activity do not vary all that much across the globe. Thus, any similarities between peoples should not surprise us—it is the *differences* (if things that appear at first to be differences actually are so) that should make us wonder.

The Maya in particular were neither peaceful contemplatives nor uncivilized "savages" rampaging meaninglessly through the countryside. Like every people, the Maya had a complex civilization created by the reflective and the maniacal together, full of kings and paupers, the virtuous and vicious, compassionate and sociopathic. Every society finds a place for its various people, and the civilizations of the Americas were no different. We know from the existing physical evidence that the Maya developed a literary and scientific culture equaling the most developed cultures in the world. They also engaged in horrific practices widespread in

Mesoamerica such as human sacrifice and violent conquest in order to advance the power and prestige of their cities. Although in general different classes of people were responsible for warfare and cultural achievement, occasionally the two converged, and we should not always expect that those capable of engaging in detailed astronomical observation, sophisticated mathematics, or the reading and creation of texts, must be distinct from those engaged in bloody conquest and the slaughter of captive men, women, and children to appease the ancestral gods. These distinct features of human nature, the lofty and the horrific, can often coexist in the same people.

One thing that the ancient peoples of the Americas did have that modern society seems to lack is a sense of their connection to the cycles of nature, to the non-human world, and to the skies. The celestial regularities, the cosmic motions, and the transformation of things within the stability of the eternal return—all of these were meaningful and grounding for the life of the peoples of the Americas. We live in an age in which we accept a different ideology—one of constant change in terms of what we call *progress*. This progress is seen as a deliberate transformation of nature, a breaking out of and a reformation of nature, bending it to our own will and our own interests. It is this, in part, that allows us to continue to neglect our destruction of the planet in the interest of advancing the economy, or economic interest. We cannot see that the nature on which we depend, including the Earth and its resources (even calling these "resources" presumes a particular conception) cannot be taken as a commodity that can be traded away in exchange for other things we value such as money. Nature is the ground of our life, and if it disappears, so do we. We cannot trade our lives away for economic gain, because we will then not exist to enjoy this economic advantage. The commodification of all aspects of our lives has led to this catastrophe in which we cannot even see the obvious truth before our eyes that we are eliminating ourselves in the name of "having a better life", which is simply madness. One who has *no* life necessarily cannot have a better life. And not only do we throw away the basis of our own lives, but also that of every other living thing on the planet. What we are doing to the Earth is akin to burning down one's own house in order to procure ashes to stencil one's walls, or the king of a city executing himself in order to remove a threat to his power.

This tenuous existence of our own world and way of life puts us in company with the Aztecs near the beginning of the Spanish conquest, with whom I close this chapter and our consideration of the astronomy of the Americas. The Aztecs had always been concerned about the continued existence of the heavens and the sun's round, in ways that the Maya never seem to have been. The practice of human sacrifice, as discussed above, grew up around the idea that the sun had to be propitiated. In order to ensure its continued rising, human hearts had to be offered to it. In times of difficulty, human sacrifices tended to rise. The sun needed to be fed. Beneath all of this we can detect a kind of paranoia about what was perceived to be the tenuous state of the world. Any day now, it could all end. The sun might abandon us, decide not to rise any longer over us, and our crops would shrivel and die, and the Earth would become a solid ball of ice. The Aztec paranoia may or may not have been justified, but today we like to think that whether or not they sacrificed humans, the sun would have still risen in the sky—there was never any real danger. Or was there?

In our own day, we stand in danger of turning the Earth not into a ball of ice, but into a spherical ball of fire, a gaseous oven, too hot for any life to survive—something akin to the fiery planet Venus. The average temperature of our planet has risen sharply, and the natural disasters we create as well as the continued dangers from the greenhouse effect will become unstoppable. Our planet could rapidly move into a state that will lead to the extinction of all terrestrial life. The tools and life of our modern world, our industrial development, and our overpopulation of the planet, depleting its biological resources—all of this contributes to our transformation of the planet and its movement toward our doom. Of course, even if we succeed in turning our planet into a second Venus, it will survive. But we will not, and neither will any of our fellow creatures. The Earth will become a planet on which life is impossible. If we are not as paranoid about this as the Aztecs were that the sun may not rise again, we should be, because our own "doomsday" is much more probable than was that of the Aztecs. Of course, expectations are often very different from reality, and while the Aztec solar apocalypse never did materialize, the end of their civilization, which they had struggled so hard to avoid, was nonetheless at hand. According to traditional post-conquest accounts, a comet appeared at the end of a 52 year cycle period on the Aztec calendar (a combination of the "short count" and vague year calendars, illustrated by the famous Aztec calendar stone of Zocala in Mexico City). Comets were often for the Aztec and other Mesoamericans, as they sometimes were for the Chinese as we will see in the next chapter, signals of impending doom. There is a depiction of the Aztec ruler Moctezuma II looking up at a massive and foreboding comet in the distance in 1512, a sign of things to come. Would the sun stop its rounds after all? But the sun did not stop. Instead, in 1517, strange ships from across the ocean appeared off the coast of the Yucatan Peninsula. Two years later they would appear off the Gulf Coast, and then into the heart of the Aztec empire itself, outside the mighty city of Tenochtitlan. The end for the Aztecs would not come in the form of eternal darkness, it would come in the form of invasion, illness, and conquest. The people of Mexico (named for this people we call the Aztec, who knew themselves as Nahua, or Mexica) finally gained their independence from the colonial power over three hundred years later, in 1821.[3] However, by then the country was much different, and certainly no longer primarily Aztec in nature. While some things remained, in culture, in the people, certainly in the land, Mexico had been Latinized. Its people followed the religion of the Spanish, built, dressed, thought, and acted much as they did. No longer were there court astronomers, sacrifices to

[3]Mexico marks its independence from the beginning rather than the end of its War of Independence 11 years earlier, on September 16, 1810. This is similar to the US practice of marking independence from the British Empire with the Declaration of Independence toward the beginning of the war in July 1776, rather than its conclusion in 1783 with the Treaty of Paris.

Quetzalcoatl, or elaborate glyphic representations of the motions of the planets. In the Maya area to the southeast, there was more survival, some Maya holding tenaciously to their culture and ways. Even down to the modern day, despite their poverty and marginalization by "mainstream" society, some Maya people follow the ways of their ancestors. Given these circumstances, it would be difficult to duplicate or extend the wondrous achievements of their ancestors in the realm of astronomy, cosmology, and philosophy. Perhaps one day the Maya will again create an autonomous and resplendent empire.

While the end came for Cahokia in the early 1400s CE, the end came for the Mesoamerican empires in the early 16th century CE. The end for stars can be either cataclysmic, violent supernovae, or (relatively) soft, simply fizzling out. Stars near the end of their lives, such as Betelgeuse and Eta Carinae gyrate unstably in the vast distance of space. Almost certainly, given the distance, these stars have become supernovae already, but the light of the magnificent explosions has not yet reached us (Betelgeuse is relatively close at under 1000 light years, while Eta Carinae is 7000 light years distant). Perhaps we will see the collapse and explosion of one or both of these massive stars in our lifetimes, or perhaps our children or grandchildren will witness this. Their skies will be lit up with a new star, a second Venus, a bright light rivaling the luminosity of the moon. It will be visible even through the light pollution of our modern cities, and the smog, dust, and haze of industrial production with which we choke ourselves and poison our children. What will these magnificent events in the sky portend for us?

Part II
The Celestial Empire

178 CE

Sometime during September, during the years of the reign of Emperor Ling in the Eastern Han dynasty, court astronomers looked out onto the night sky and saw the emergence of a long, brightening object, one that looked like a central star with streaming bristles emerging from it, stretching for many degrees of the arc of the sky. It appeared somewhere in the vicinity of what we know as the constellation of Hercules, and as the days went by, its brightness increased, becoming a bright white and red streak cutting across a vast swath of the sky. The court astronomers tracked the star for about 80 days, as it brightened then slowly faded, and finally disappeared from sight. It was a "broom star"—they recorded it and moved on. Though had seen such things many times before, this was a particularly magnificent appearance. This broom star vastly outshined and outreached anything the court astronomers had ever seen. Seeing such an amazing and unprecedented celestial display must have caused some of those astronomers, as well as other members of society, to wonder. The empire seemed to be in decline, the powers of eunuchs and members of elite families often getting in the way or eclipsing that of the emperor himself. Numerous scholars were writing of the decline of morality, the intellectual stagnation of the court, and its increasingly weak and reactive style of governing.

They had to wonder: was this the end?

Chapter 4
The Chinese World

The Celestial Sphere

One of the major differences between Chinese astronomy and that of the Americas is that the Chinese were focused on the *sphere* of the heavens rather than on the horizon. This focus on the celestial sphere should be understood in a modeling sense. Though there were, as we will see below, a number of different cosmological views in early China, the view that the universe was literally in a spherical shape was not a popular one, unlike in the ancient West.

The armillary sphere was an astronomical instrument well known in many stargazing civilizations. In its basics, the tool is simple and elegant. It operates using the fundamental idea of the celestial sphere, which is a projection of the sky as a sphere surrounding the Earth. Not only the visible night sky, but the entirety of the visible objects in the heavens. The purpose the tool is to map the positions of important features of the sky, as well as chart *motions* of celestial objects, in a sphere that can fit atop a desk, or in smaller versions even in the palm of one's hand. The armillary sphere thus has functions that a simple celestial globe, which is a projection of the celestial sphere without the movable rings of the armillary sphere, does not (though the early Chinese used the celestial globe as well). The armillary sphere allows for *dynamic* measurement—and as we will see, the idea of change and transformation is a key to understanding early Chinese astronomy, and Chinese thought in general.

With the armillary sphere, one can tell the time, chart any celestial phenomena, tell the season, and plenty of other useful features that would have been invaluable to almost any person throughout most of human history, especially farmers, navigators, and cartographers (among others).

Many stargazing cultures used versions of the armillary sphere, even though the character of their astronomical thought was sometimes very different. For the Greeks, and much of the Western world until the modern era, astronomy was based on the ecliptic—that is, the apparent circle through which the sun travels, and most

A. McLeod, *Astronomy in the Ancient World*,
Historical & Cultural Astronomy, DOI 10.1007/978-3-319-23600-1_4

of the planets stay close to. Of course, the ecliptic is actually defined by the orbit of our own planet around the sun—as we know, it is we who move around it, rather than vice versa. Ecliptic-based astronomy is mainly motivated by a primary concern with following the planets. This is because it is the planets that roughly lie along and follow a path along the plane of the ecliptic.

Chinese astronomy, however (from the most ancient times onward) was *equatorial* in nature. In this way, contemporary astronomy shares more in common with early Chinese ways of viewing and charting the sky than it does with Greek ways, and Western ways in general all the way to the inauguration of new and more advanced equatorial methods by Tycho Brahe in the 16th century. Rather than following the plane (roughly) of the planets and the earth's orbit, the Chinese based astronomy on movements around the celestial pole.

Because the earth is tilted on its axis from the plane of the ecliptic, the plane in which we orbit is not the same as the plane in which we spin daily. The motion of the stars in our sky and our sun, of course, are based on *both* ecliptic and equatorial motion. But each is responsible for a different aspect of this motion. Equatorial motion is responsible for the daily motion of the sun and the stars, along the plane of the equator, with the north and south pole line as the axis. On the other end, it is ecliptic motion that is responsible for the annual motion of the sun through the zodiac, for example, which the Greeks were far more concerned with than the Chinese, who were more interested in the location of the moon. For the ecliptic astronomer, planetary motion and zodiacal movement of the sun will generally be the primary concerns, while for the equatorial astronomer (at least in periods before the modern age), circumpolar motion, lunar motion and position, and locating objects will be the main concerns.

In this chapter, we will look at length at the unique features of Chinese equatorial astronomy, but I want here to mention at least a few things that distinguish it from the ecliptic astronomy, and that we in the contemporary world might take for granted, mainly because we today operate using the Chinese system, and not the Greek. While the modern West often likes to believe it is completely the descendent of Greek culture, our astronomical system (and many other things we do not realize or sometimes admit) was born in China, not in Greece or anywhere else in Europe.

If we take the line running between two poles as the axis on which the earth spins, the equator is defined by the plane of lines running perpendicular to this line, with origins at the point halfway between the two poles. We can extend each of these, the plane of the equator and the axis line, outward, so that they go beyond the earth and onto the celestial sphere. Thus, the equator of the celestial sphere is an arc along the plane of the equator, halfway between the two poles of the sky. For observers in the northern hemisphere, such as those in China, only the north pole will be visible. Today, the star Polaris inhabits a location roughly at the pole, and this is why it has the popular designation "the North Star" or "the Pole Star". For any observer, the location of the north star in the night sky will be exactly the same as one's longitudinal position, or elevation from the equator. Note that this will be true for observers *anywhere on Earth*. So, for example, if you are in Hong Kong, at about 22.5°N, the north star will appear in the north at 22.5° elevation from the northern horizon. If you are in Beijing, at about 40°N, it will appear at 40° elevation from the northern horizon. Thus the elevation of the north star in the sky (if you're in the northern hemisphere and can see it) will always tell you at what latitude of elevation you are.

This explains part of the usefulness of the north star for navigation in the premodern world. Notice that the position to latitude correlation just mention holds for *any* position in the world, even southerly locations in which the north pole cannot be seen! How is this? Our system of latitude is based on angle of elevation from the plane of the equator. The equator is thus zero latitude. Any point north of the equator has a higher latitude, with the numbers increasing from 0 to 90. The north pole itself is 90° north latitude, as it is at this point that the angle of elevation from the plane of the equator is 90°. There is a very different 90° elevation from the plane of the equator—namely, that of the south pole. Thus, there are varying latitudes of elevation from 0 to 90 as one goes south from the equator as well, corresponding to the 90° between the equator and the south pole. Thus, when we determine longitude, we have to be clear about whether we are talking about north or south, as this will give us very different places. 90°N is the north pole, 90°S is the middle of Antarctica. Sometimes N and S are used to distinguish the two, while sometimes one fines + and −, which I personally am not a big fan of, as it suggests the northern orientation is the "right" one. The positive and negative designations will be helpful here to make the point about the north star. If one is at a latitude of −15.3, the correlation with the north star still stands, even though one cannot see it. In fact, the correlation in part *explains* why one can't see the north star, because at this latitude the north star is 15.3° *below* the northern horizon. Likewise at 47°S—the north star is 47° below the horizon. And when one is standing at the south pole, the north star is straight beneath your feet (Fig. 4.1).

In Chinese astronomy, as we will see, the north pole has enormous significance, and it is likely primarily this fact that accounts for its equatorial nature. But I must mention here one further feature of this system: its excellence for creation of coordinate systems, in comparison with ecliptic astronomy. Because it is this in part that explains why the Chinese became such masterful observers.

There are two main systems used for coordinate systems in observation—one using altitude and azimuth, based on the horizon (we might call this *horizontal* astronomy), and we saw above that many of the people of the Americas practiced this kind of astronomy, due to their concern with risings and settings, and one using ascension and right elevation, based on the celestial equator. The Chinese, as well as modern astronomers, used the latter. While the latter (often called alt-az) system can be extremely useful for observations of the horizon and the zenith (which is just what astronomers in the Americas were most interested in), its usefulness becomes much less when we shift our concern to objects and events happening other higher elevations in the sky, specifically the polar regions. In addition, tracking the movements of objects across the sky becomes much more of a burden with an alt-az system, as the daily motion of the planet will change the coordinates of any given object as the planet turns. The reason for this is that the alt-az system is based on the sky as visible dome, rather than the celestial sphere base of the equatorial system.

Fig. 4.1 The equatorial sundial can be helpful in understanding the celestial sphere. This sundial is basically *half* of a celestial sphere, with the horizontal ring following the celestial equator, and the vertical ring following the meridian (north-south line). The *top* of the vertical ring represents the celestial north pole, and is tilted to the elevation of the northern pole (the *bottom* of the ring points to the south pole, and in the southern hemisphere the southern part of the ring would point toward the sky). The elevation of the pole from the horizon is the same as the elevation in latitude of the location from the earth's equator. Chicago sits at about latitude 42°N, and thus the northern tip of the north-south ring of this sundial, in front of the Adler Planetarium in Chicago, is elevated 42° from the horizon. The angle between the gnomon line (connecting the two poles of the meridian ring) and the horizon line is then 42°. Photo by the author

Following and accounting for motion around the equator, the apparent motion of the stars and everything in the universe outside of our solar system, is made most manageable on an equatorial system. And it is just the Chinese who adopt such a system of practice. The armillary sphere is a brilliant illustration of this system, a tool for observation that moves as the Earth itself, and that can be used to spot and determine equatorial position for any object in the sky. The armillary sphere of the Chinese, based on the celestial equator, moved in almost the exact same way as our modern equatorial telescope mounts, which are their descendents. The resting position for the sphere, as for our modern mounts, is not for the sights (or the scope) to be pointed straight forward at a point on the horizon, as one might expect. Rather, it is pointed directly at the North Star, which has coordinates 0, 90. Direct north, zero point along the equator, and 90° north from the equator in elevation (or right ascension). Of course, one complication of an equatorial system, but one that can be easily overcome, is that just where 0, 90 is in any given place will be different, and thus the coordinates of any given object will give us a different actual place in the dome of our sky depending on our latitude. This turns out to be a *virtue* of the system in comparison with the alt-az system, rather than a vice. The reason the 0, 90 point of the north pole is the most natural resting point for the equatorial sight or scope is that this position can be determined in any given northerly location very easily (it's a bit tougher for those in the southern hemisphere, who don't have a pole star). All we need to do is align the sights or scope to fix on the north pole, which is the only object in the sky that does not move its location, and thus gives us a fixed reference point (Fig. 4.2).

The armillary sphere, like the equatorial mount, then has two directions of motion. Left and right along the plane of the equator, and up and down along right ascension lines, or elevated relative to the equator. Notice that the motion is not centered on the sights (or the scope), but on the equator, and the initial position of the sight is 90° elevated to the equator, fixed on the pole. Thus, from the initial position, if one moves the sphere (or mount) along the equatorial plane, the sight will not move from the North Star, but simply rotate around. This is the only spot in the sky in which this will happen—just as something similar would happen in an alt-az system if you pointed a sight straight up in the air at the zenith and then rotated around toward every point in the azimuth. Because the north pole is at the "top" in the equatorial system, we find this effect. The other plane of motion, right ascension, will then bring the sight closer to the equator from the 90° initial point (you cannot get further from the equator in right ascension than 90°, as when you pass the 90° point, you are closer to the equator in another direction, and thus have a less than 90° angle of elevation in that direction. RA is always calculated in terms

Fig. 4.2 Chinese celestial globe, created in the early 19th century, in the collection of Adler Planetarium, Chicago, IL. Photo by the author

of *closest* angle of elevation with the equator. There will always be *two*, if we base this on the 180° between the two equators (we can neglect the elevations *below* the equator, which are not relevant for astronomical purposes, either on alt-az or equatorial systems) (Fig. 4.3).

As mentioned, the genius of this system is many fold. Not only is there a fixed point at the pole to set initial position, but based on the pole and circumpolar stars, there is a way to give every object in the sky a fixed declination and right ascension, so that their coordinates are independent of the motion of the sky, and of the motions of the planets, instead fixed on lines from the non-moving north pole. In alt-az systems, given that the zenith and thus the 0, 90 point does not point to a fixed position, objects will have different coordinates throughout the year, given the annual progression of the earth and thus the different tilts of our axis and equator relative to the sun and stars. What occupies the zenith tonight at 9 will not occupy it at 9 two months from now. Even worse, what occupies the zenith at 9 tonight will not occupy it at 11 tonight, unless we are at the north or south pole, where it is just the unmoving pole point that is our zenith. Thus, we can see the clear advantage of equatorial astronomy.

The armillary sphere has one additional feature, beyond observation of the stars. As based on the natural motions of the planet and mirroring the celestial sphere,

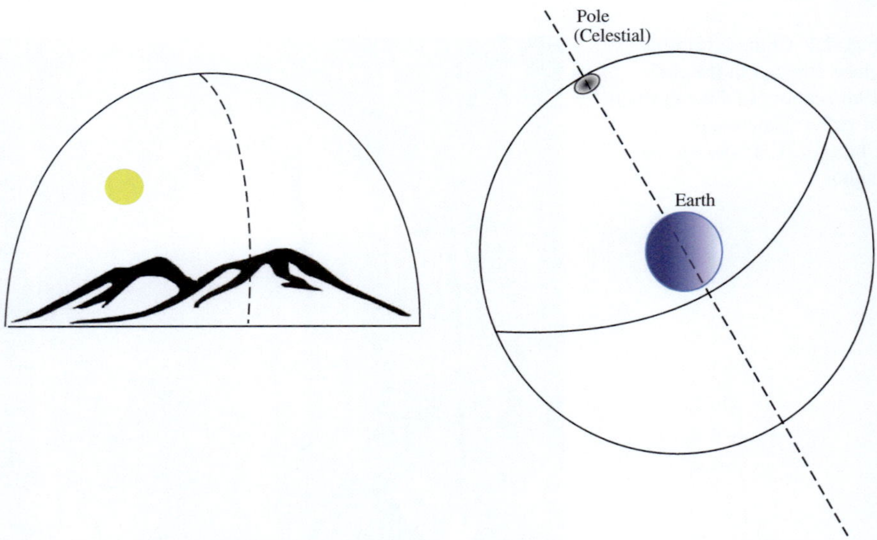

Fig. 4.3 Alt-az coordinates (left) are made using the horizon and zenith as references. An object is located on the dome of the sky by taking two coordinates. Altitude is determined by the angle of elevation of the object to the horizon (where 90 degrees is the limit, at the zenith). Azimuth is determined by the degree of arc along the horizontal from north of the point on the horizon above which the object is located. Since there are 360 degrees in a circle, this is also the limit of measurement (360 degrees = 0 degrees). Coordinates in a polar-equatorial system are based on the celestial sphere, in particular the pole and equator. It is somewhat like the alt-az system in that it uses elevation for one coordinate and position along a ring for the other, but the place of the zenith in the alt-az system is replaced with the pole, and the place of the horizon is replaced with the equator. (Illustration by the author)

it also then mirrors the movements of the Earth itself. We can see the celestial sphere as a projection of the spherical image and motion of the Earth onto the background of the stars. This is the "sphere" viewed from "inside". If we could get outside the sphere and observe its operation as such, it would exactly mirror that of the Earth. The genius astronomer, poet, philosopher, and statesman Zhang Heng, who lived during the Eastern Han Dynasty, recognizing this parallel, considered that he might create a replica of the motion of the earth if only there were some way for him to get the sphere to move in sync with the motions of the earth—that is, to *keep time* with the Earth's motions. How could this be done? If the sphere could be mechanized so as to rotate completely once every 24 h, as well as to rotate back and forth on its axis between its degrees of tilt once every year, this would be to create a mechanically powered replica of the earth, duplicating its motion.

For this task, Zhang employed the most accurate known timekeeper of the period, the *water powered armillary sphere*—basically a motorized armillary sphere powered by falling water. Zhang realized that an accurate model of the heavens could be constructed if an armillary sphere could be kept in motion closely approximating that of the diurnal motion of the celestial sphere. Zhang had the further ingenious idea to *power* the sphere using a water clock, in which time is kept by a regular flowing of water, thus killing two birds with one stone. This innovation (as well as a number of Zhang's other inventions) combined two well-known tools. The water clock was known in China long before Zhang Heng, and it was also used in ancient Greece, where it was known as a *clepsydra*. Its antiquity in world history goes back to ancient Egypt, more than a millennium before Zhang Heng's time. The combination of the armillary sphere and the water clock was a superb innovation that represented in working form a Chinese conception of the cosmos, even in the working of the machine. The element of water and its flowing linked directly to time, which operated the model of the cosmos. The *wu xing* (five processes) and the connection between the cosmos and humanity could be recognized through this machine. Thus it had more than one purpose, just as did many of the astronomical instruments of ancient China and the other cultures discussed in the book. Its purpose was tied to the Chinese conceptions of the celestial order, which revolved (literally) around the pole.

Understanding the role of circumpolar motion in early Chinese astronomy is key to understanding how to read the augury of the 178 comet. But to see the importance of the pole, we first need to understand the nature of a few unique features of Chinese astronomy.

By the 1st century CE in the Chinese states, astronomical prediction was a well-established practice. A devoted professional class of sky watchers charted every motion and phenomena of the sky in the finest detail. The Chinese empire had perfected the art of astronomical observation early on in the years BCE, and would remain the most skilled of observers well into the modern period, only losing their dominance after the age of the telescope and Tycho Brahe's revolution in astronomical observation, in part made possible due to knowledge from early China.

Almost all unusual celestial phenomena recorded before the 1st century BCE, such as novae, supernovae, comets, and eclipses, were recorded in Chinese records.[1] In these early years, Chinese accounts are often the only existing accounts of these phenomena. In years after the 1st c. BCE, Chinese observations of these objects are often first and longest. The Chinese court astronomers found new comets first, and saw them disappear last. This is the case for most objects up until the 16th and 17th centuries. There are likely two reasons for this. First, astronomy in China was valued as an art of central importance, and thus there was a gainfully employed and dedicated staff of court astronomers at all times, whose purpose was to observe the skies and make predictions based on their observations. Second, because of this focus, an enormous amount of intellectual energy was spent on observational astronomy, and the Chinese developed fantastic tools to aid in their observation of the skies, tools that would be unmatched elsewhere in the world for many years. In short, the Chinese dominated observational astronomy for most of the history of humanity simply because they *cared* about it, they valued it in a way few other societies did.

In the earliest scholarly texts in China, we hear about the intrinsic connection between humanity and nature, a connection that will later become the basis of much of the philosophical and scientific thought in the beginnings of empire.[2] The workings of the natural world are in harmony with those of the human realm, in such a way that phenomena discernible in nature can help to presage events that will happen in human society. Thus, the idea of divination based on natural phenomena arose early in Chinese society. And, when we think of which aspects of nature might best serve as a measuring stick for human events, we will inevitably be drawn to those features of nature most stable through time, that yet offer enough changes for us to make sense of the occasional chaos of the human world. A parallel to human life: much of it is regular, cyclical, and dependable, but occasionally events happen that appear to us unprecedented, chaotic, out of nowhere, events that don't easily fit within a tidy theory or system. The sky is a nearly perfect parallel for this-perhaps the best we can find in nature. The sun, the stars, and the planets maintain their regular motion seemingly eternally. It is only in very recent years, and with the most sophisticated of technology, that humans have been able to discern that the objects of the heavens themselves evolve, move, develop, and eventually die as well.

At the same time, there are often unexpected and seemingly random events in the skies that seem to violate this sacred order, that thumb their noses at the eternal regularities of the celestial motions. These phenomena the Chinese knew well, and

[1]Many of these records, along with some from Japan and Korea, have been compiled from their numerous sources in (Xu, et al. 2000).

[2]Well known texts such as the *Xunzi*, written in the Warring States Period before Qin "unification", *Lushi Chunqiu*, and early Han texts such as *Huainanzi* and *Chunqiu Fanlu*, among others, stress this connection. Some early texts such as the "Daoist" texts *Daodejing* and *Zhuangzi* may also be making this "naturalistic" connection. Janghee Lee discusses early Chinese naturalist views in (Lee 2005), in which she resists a reading of Xunzi as "naturalist". As is well known in contemporary philosophical circles, of course, "naturalism" is a tricky concept, and it is not always clear what one means by it.

documented closely. Such phenomena as the "guest star" (nova or supernova), "sparkling star" or "broom star" (comet), for example. They appear as anomalies in an otherwise predictable system. Random and temporary celestial flare-ups that get in the way of an orderly explanation of the motions of the heavens. Today we understand that these phenomena too respect physical laws and fit into a system that, although orderly, is far more complex than we ever could have conceived in the ancient world. Even comets, the proverbial galactic wanderers, follow orbits, respect the laws of gravity, and can be thus predicted (given enough prior information about their makeup, etc.). But to pre-modern people around the globe, such celestial phenomena could not be explained by even the best models of the workings of the heavens, and could only be seen as anomalous.

Ptolemy's *Almagest*, in the West, asserted the theory of the celestial spheres, with Earth at their center, and included tables to aid in prediction of the motions of the planets. There was no room in this system for theoretically unpredictable events like comets and novae. Yet such events happened in the night skies–and often. Current estimates put the average number of naked-eye visible comets during any given year at around 18. Not to mention the number of solar and lunar eclipses, novae and supernovae that must be added to this list. How did the Western astronomers make sense of these? In a classic example of "saving the appearances" along with one's pet system, phenomena such as comets or novae were written off as meteorological, or *atmospheric*, phenomena. Since the Aristotelian/Ptolemaic system could not make sense of comets and novae, they must simply not be astronomical objects. They must have their origins in the air, similar to clouds. We recognize a kind of chaos in Earth's atmosphere, in wind, clouds, storms, etc. that is obvious. So it must be the case that the seemingly chaotic comets and novae are likewise atmospheric events. Explanatory system saved.

For this reason, even when there were celestial events Western astronomers clearly must have seen, they were often not recorded, instead dismissed in the same way a modern astronomer would take no notice of the clouds floating across the field of view of their telescope and blocking out their view of Jupiter. This was not the case for the Chinese. They made room in their theories of the universe and the heavens for change and for unpredictability. The parallel between nature and the human realm gives us a perfect way to make sense of this. Human life, although following regular paths, is also unpredictable and chaotic occasionally. Our lives are marked by this strange combination of the orderly and the chaotic, the familiar and the foreign, the predictable and the random. Unlike the Western view, the Chinese conception of the natural parallel between humanity and the rest of nature allowed for the other parts of nature to also act in this way. And indeed, this seems to be just what we observe in the skies.

Because of this, early Chinese astronomers took unpredictable and transient events like the appearances of comets or novae to foretell similar changes in the human realm. Given the connection between humanity and nature, a disruption or change in the sky could be expected to correspond to such a disruption in human society as well, either for good or evil. The main purpose of the professional astronomers in Chinese society was to observe and interpret these signs, to read the message of the stars and record (mainly for the court and the emperor) their meanings. The Chinese astronomers thus had an interesting combination of talents, what we might consider today part scientist and part witch-doctor. They had to be excellent observers, and indeed they developed observational tools and systems that far eclipsed all others in the world for most of human history. But they also had to be shamans or diviners, astrologers–they had to understand how to discern the significance of the appearance of a certain comet, for example, for the human world (specifically for the government by which they were employed). They had to understand the influences of certain celestial objects over certain aspects of society and persons, to understand how the chaotic phenomena were attached to human events and what human events they foretold. In ancient China, as in all other places and times, such divination was as much an art form as much as anything else.

To call it an art form should not be taken as a dismissal of the practice. Indeed, the Chinese had a sense of the heavens and understanding the heavens as art that we in the contemporary world have largely lost. This may be part of the reason for the rapidly diminishing sense in our own culture of the value of the night sky, of astronomy in general. As with other arts, the significance of the astronomical art is in its connection to our sense of meaning, our thoughts and ideals, our reflection on the world in general. One of the central tenets of "science" as it is conceived in our own society is that it is based in predictability, repeatability, and ultimately systematicity. Art that is systematic, however, is dead. Art in a large sense represents the unpredictable, the unconstrained and unsystematized aspects of our lives and nature.

Chinese astronomical divination was an art in more ways than one. Interpreting the meanings of certain celestial phenomena, of course, was not as easy as consulting a table of meanings in a giant encyclopedia, or reading the answer from a giant Magic 8-Ball. Meanings had to be interpreted, gleaned from the skies by those familiar with them. If a comet were to appear, its location in the sky, its brightness, its length of observability–all of these were relevant to its meaning for the human realm. There were no quick and easy rules (in fact no rules at all!) by which to determine what such an event signified. To do so, one had to develop the skill of an artist, an internal feeling, a "nose" for it. Just as with painting or drawing–one cannot follow a list of rules to generate an excellent painting. Intuition, skill, and a certain amount of license are necessary. Regularity, duplication, and systematization, while they may be the birth of science, are the death of art.

Astronomy and the Fall of the "New" Dynasty

It was toward the end of the Eastern Han dynasty that the great comet of 178 appeared. By this time there had already been three other "dynasties" in Chinese history, and in each of these similar celestial events were occasionally recorded, although perhaps none quite as spectacular as the comet of 178. The first dynasty, created by the ruler of the state of Qin, lasted a mere 15 years, barely lasting past the death of its founder. The second dynasty was the first of the great dynasties. The (Western) Han rose from the ashes of the Qin, created by the rebel leader Liu Bang, whose family would become central to the history of China for many years to come. The dynasty of the Liu family was to last for about 200 years, before a crisis in the very early years CE brought the Western Han to an end. A third dynasty, often overlooked in the histories as such, arose in this period. It was to prove as short lived as that of the Qin. This Xin ("new") dynasty (9–23 CE) also suffered in the later histories in part because the dynasty following it saw itself as a restoration of the Han, even though it was created by a different branch of the Liu family than that of Liu Bang. The second (Eastern) Han dynasty would last for another 200 years. The history of early Chinese empire began to take on familiar recurrence—a short and influential dynasty followed by a strong and long-lived dynasty named Han and controlled by a Liu. Both the Qin and Xin dynasties were much more influential than many Han historians gave them credit for being,[3] and both have been effectively vilified in Chinese history. In the story of the founder and only emperor of the Xin dynasty, Wang Mang, we can see a small part of the role of astronomy in early Chinese life. Whether or not the astronomical portents of doom for dynasties like the Xin were invented or intentionally taken out of context by Han scholars, the attributed signs to the appearances of celestial events was taken seriously. The sky would signal the end for the Eastern Han as well, in time.

Wang Mang, emperor of the short-lived Xin dynasty, was one deeply concerned about astronomical prognostication and intellectual legitimacy in general, as many newly empowered political regimes were (and are). By the time the doomed Xin was coming to a close in the years surrounding 23 CE, divination and prognostications based on celestial phenomena had a deep and revered history in Chinese government. Astronomers, or, we might say *astrologers* (the distinction is one that breaks down in east as well as west, and one we will have to consider again later in connection with Western thought) studied the sky closely, with tools as precise as any available in the ancient world and even until very recently in human history. Their observations were not only scientific, but had a political import as well. And

[3]It is understandable that Han scholars and other representatives would want to diminish the accomplishments of the regimes they displaced, but scholars have shown that both Han dynasties retained many of the innovations of their predecessors. Sanft (2014) and many others discuss the Qin influence on Western Han. Loewe (2006, xiv) discusses the Qin and Xin influences on the two Han dynasties: "Paradoxically enough, it was the two emperors who have been subject to the most severe hatred for two thousand years whose regimes left a lasting heritage which their successors were happy to accept and to apply to their own problems. It was in the reign of the first of the Qin emperors that imperial institutions took shape; the succeeding emperors of Western Han adopted them as a means of governing their realm, and parts of that heritage such as the division of the land into counties may still survive today. Wang Mang set himself up as emperor of a dynasty known as Xin (New) from 9 to 23 CE. He gave out that he was following in the footsteps of the kings of Zhou, and that he was adhering to their traditions, and he re-named some of the existing institutions of government to accord with what he believed to have been their practice. Despite rejecting such nominal changes, the emperors of Eastern Han who followed could hardly jettison the moral support implicit in Wang Mang's claims and pretensions."

even beyond this, the observations astronomers made had significance for the entire world, for all under heaven. The motions and changes in the sky were seen as connected intrinsically to the human world. A supernova or a comet (depending on how interpreted) could be the harbinger of natural disaster such as flood, famine, or earthquakes. Alternatively, it could be the sign of coming years of prosperity. It could signal the imminent end of a kingdom or empire, the birth of a new one, or the strength of an existing one. For one skilled in reading the skies, they held the key to understanding the human past, present, and future.

Wang Mang had risen to power in 9 CE, after the power of the Western Han dynasty faded through the increasing weakness of the central court and the increasing power of landholding families, most particularly the Wang family. While the earliest years of Wang's rule saw a number of economic and intellectual innovations and a formation of government around Confucian moral principles,[4] natural disasters and rebellions quickly began to undo Wang's dynasty.

The Eastern Han historian Ban Gu recounted a prediction of Wang Mang's fall by court astronomers of his Xin dynasty. Or at least after the fact, with the benefit of hindsight, it seemed a prediction. It is hard to imagine court astronomers being so bold and unconcerned with their own tenuous fate as to predict the downfall of their court. Unfavorable portents of this magnitude could be grounds for execution.

Ban's *Later Han History* recounts the observation of a comet in November of 22 CE coinciding with disastrous events that would eventually lead to the premature end of Wang Mang's reign and his newly established dynasty. Later history would view Wang's short reign as a kind of "interregnum" period between the two Han dynasties, an awkward anomaly in an otherwise orderly progression of dynastic houses through the years. Of course, this picture is just as fictional as the official view of Wang's rule. Dynasties did not cede to one another or progress in natural manner. Every change of dynasty was inherently usurpation, a conquering and destruction of one ruling house by a challenger who then gained the ability to write his own history. It always paid to accord oneself legitimacy by claiming some kind of cosmic continuity with past "dynasties" in the victor's history, which is, of course, all we have. Interestingly enough, the Later (or Eastern) Han had no desire to connect themselves to Wang Mang, and thus constructed their claims to legitimacy based on the earlier Han. Eastern Han scholars emphasized their connection to the Western Han, such that even today we sometimes think of the Han Dynasty as a single imperial dynasty spanning from 206 BCE to 221 CE.

A few comets had been noted during Wang Mang's reign, although the only extant source we have for the history of the Xin Dynasty is from the Later Han author Ban Gu, whose account is (of necessity) biased against Wang, given that Ban is an agent of the dynasty that displaced Wang.[5] The short descriptions we see

---

[4]Even authors hostile to Wang's regime, such as the Han scholar Ban Gu, writing in the *Han shu* ("History of Han") recognized Wang's praiseworthy attempts to organize government under moral principles. This idea is further discussed in Thomsen 1988.

[5]We do not even know the reign name Wang took on his rise to become emperor, as Ban Gu does not record this, probably so as not to accord Wang legitimacy.

in Ban's *Han Shu* (Book/History of the Han) of comets during the reign of Wang Mang are suggestive of inauspicious omens, and if we take the fate of Wang and his dynasty as indication, they indeed *were* bad omens. Generally, a regime in trouble will try to reduce discussion of possible negative signs, including comets, supernovae, and other signs (including natural disasters, which are, unlike celestial phenomena, *always* bad omens). If a regime is flourishing, it has little to worry about when unusual celestial phenomena appear, because such are clearly an indication of the favor of heaven/nature. But it is easy and natural for the people to connect such phenomena, in the case of a faltering regime, to the displeasure of heaven. Thus, the astronomers controlled (or at least constrained through their employment) by a court had reason to downplay any event they read as having negative implications, and one way to do this would be simply not to mention its divinational import. A comet appeared on such-and-such a day. The end. We see lots of this in the Wang Mang memorial of the *Han Shu*, but it is unclear whether Wang Mang's own astronomers documented the events in this way, or whether this is a literary flourish on the part of Ban Gu, to indicate the correctness of the fall of Wang's dynasty, and the establishment of the glorious dynasty of his own employers, the Later Han.

While Wang perhaps was not the most effective ruler, or it may alternatively have been the case that natural events beyond his control, such as persistent droughts during his reign, ultimately led to his undoing, even the harshest critic such as Ban Gu could not deny that Wang displayed an admirable concern for intellectual development. In the vein of his predecessors in the late Warring States and early Han, Wang called councils of scholars together to produce works of philosophy, astronomy, and history. There is an account in the *Han Shu* of his organization of just such a council.[6]

Despite his concern with intellectual production, we do not today have any written works that we know were produced during the Xin dynasty. Part of the reason for this may be the influence of the Later Han. If any of the works produced by Wang Mang's councils indeed survived into the Later Han, they would likely have been either coopted by the Han or destroyed by it. Like any new regime, the Later Han would not have wanted materials presenting the ideology of the previous one infecting the minds of the people. So it may be the case that the works of the Xin dynasty scholars are lost, and it also may be the case that we have parts of these works today, subsumed within Later Han works. Another alternative is that, given how short the Xin dynasty turned out to be, the scholars collected by Wang Mang simply did not have the time to produce their intended comprehensive works before the fledgling empire imploded. Though admittedly it is hard to avoid the belief that they must have created *something*.

Wang Mang's rule came to an end rather ignominiously, after a short rule of deepening crisis. The final turn came, according to the author of the History of Han, in the last year of his reign, and his life, the summer of 23 CE. One of the servants

[6]Dubs ….

to a minister of the government who dabbled in astronomy—an ancient amateur astrologer, if you will—told his master that a comet that recently appeared near the celestial pole, the emperor's residence, augured a restoration of the Han dynasty. There would be a return to power of the Liu family, the founders and maintainers of the Han, the dynasty that Wang had displaced. Thereafter, the master, Wang She, hatched a plot with members of the Liu family to bring down Wang Mang. What better plan than to be on the side of the faction destined by the heavens to prevail?

Liu Xin, the head conspirator of the displaced family that had controlled the Han dynasty, would not allow the plot to be launched until the appearance of Venus, which had disappeared in its transition from morning to evening star, and would soon emerge again. We see here a similarity between the Chinese and the Maya conceptions of Venus as important patron of war. The influence of Venus could make the difference between victory and defeat, life and death. Unfortunately, the delay in action led to the discovery of the plot by Wang Mang (time always allows more chances for rumor to travel), and in the end a number of the conspirators were executed, while Liu Xin and Wang She were forced to commit suicide. And much of this was due simply to the timely appearance of a comet.

Wang Mang was not to outlast his assassins for long. A number of rebellions broke out, in the west as well as in the south, hemming in Wang in the capital region in the northeast. These revolts took up the cause of the Liu family and the Han dynasty restoration as their aims, and they were ultimately a force too great for Wang to resist. While Wang She, taking the augury of his servant to be about Liu Xiu and himself, had been proven wrong, the prophecy would ultimately be vindicated—the Liu family would return to power, and form a second Han dynasty. Wang Mang in the days to come suffered defeat after defeat at the hands of the rebels, who finally arrived at the doorstep of the capital itself, Chang'an. They easily penetrated its defenses, after which they desecrated the tombs of Wang's family members—an important first step in wiping out his influence and memory—and then, finding Wang Mang holed up in his palace, killed him, and put an end to the "New" dynasty, which had lasted a mere 14 years. There was now a new dynasty of the Han.

Although this new dynasty called itself 'Han', as had the dynasty that had been displaced by Wang Mang, to see it as the same dynasty as the previous Han would be a mistake. The founders of this new "Han" were indeed Liu, but they were only tenuously linked to the old Liu family that had formed and ruled the former Han dynasty, and would not have been considered members of the imperial lineage had the old dynasty survived. Nonetheless, this new Liu family took advantage of its connection to the Lius of old, and used this connection as part of a powerful justification for their rule—that they represented a restoration of the Han dynasty, rather than just another "new" dynasty like the Xin. China had tried something "new", and it had been a disaster. The people wanted a return to the old, to the halcyon days before droughts, mismanagement, and revolution. Of course, human memory can sometimes be very short. People had forgotten just how bad things were in the closing years of the Han, a situation which had itself led to the end of that dynasty and the rise of Wang Mang. They only saw as far as the last crisis. And "restoring" the Han was widely seen as the right answer.

155 years later, this second Han dynasty was itself on the ropes. And just as with Wang Mang, the message as to its survival or demise, everyone knew, was to be found in the skies.

Early Chinese Cosmology

In China, as in the West and elsewhere in the world, there was no single agreed upon conception of the organization of the cosmos. There were competing theories, backed by different scholars and different schools, which enjoyed more or less success depending on the time, influences with rulers, and other political and intellectual factors. Of course, some of these cosmological theories gained greater influence in Chinese culture in the long run than others, just as philosophical schools like Confucianism gained a greater long-term influence than schools such as Mohism or Legalism.

Although there was divergence on the view of the nature and structure of the universe, there were also a number of regularities concerning the sky lore and mythology in Chinese astronomy. Every culture and people has its sky lore, connected with religious or philosophical myths making sense of the human world, or often the creation of humanity (as in the Maya story of the twins Hun Ahaw and Yax Balam, and their trip to Xibalba). The ancient Greek sky lore is familiar to us today, as this is the most well-known and often-used system in the contemporary West. When we look into the sky, we see the Belt of Orion the hunter, the prophet maiden Cassiopeia, Pegasus, Andromeda, and so on. There are mythical stories behind each of these constellations, and this is also the same for every other tradition in which there is sky lore. All societies have drawn constellations with the bright stars, to help them navigate and recognize objects in the night sky. These constellations have not, of course, always been the same, with a few exceptions. It seems to be an almost universal feature of cultures to see the circumpolar star cluster that many refer to as the Big Dipper as a dipper of some sort. (Though the ancient Greek imagery we use in official designation is of a large bear—Ursa Major, in distinction to the little bear—Ursa Minor, which many also refer to as the Little Dipper, whose handle ends with the star we mark as the pole star, Polaris) This is common to so many cultures that some have hypothesized that the identification of these stars with a dipper may have originated with the very earliest human communities, arising in our common places of origin in Africa and the Middle East. This would be an amazing fact if true—it would mean that when we look up at the night sky and see there a dipper, we are expressing a shared cultural connection with our earliest human ancestors.

According to the Chinese view, the circumpolar stars represent the palace surrounding the emperor, who is the pole star, and the various members of the celestial bureaucracy. Indeed, the Chinese saw the night sky as a mirror of the empire, and saw the empire as a mirror of the sky, on earth. The sky was linked to *tian* (heaven, nature), and the empire had the authority of *tian*. It was the identification of the majesty and enormity of the night sky with the all-pervasive and mighty Chinese empire (ruled by numerous dynasties throughout its history all the way up to the Republican period of the 20th century, and the end of the Qing—perhaps ironically, China's final dynastic empire was one that was not Chinese, but Manchu) that led to the Chinese reference to its empire as *The Celestial Empire*. The empire was not just a representative of heaven—it was the heavens on earth (Fig. 4.4).

The image of celestial empire was central to much of the star lore of the Chinese, especially that concerning the pole and the circumpolar stars. We should not assume, however, that the pole was the star we know as Polaris—for much of Chinese history, there would have been a different star closer to the celestial pole, or no star at all. This is because of the effect of precession, as discussed above. Today, Polaris is the closest visible star to the pole and so serves as the pole star, but the "Pole Star" referred to in ancient Chinese texts reaching back to the 3rd century

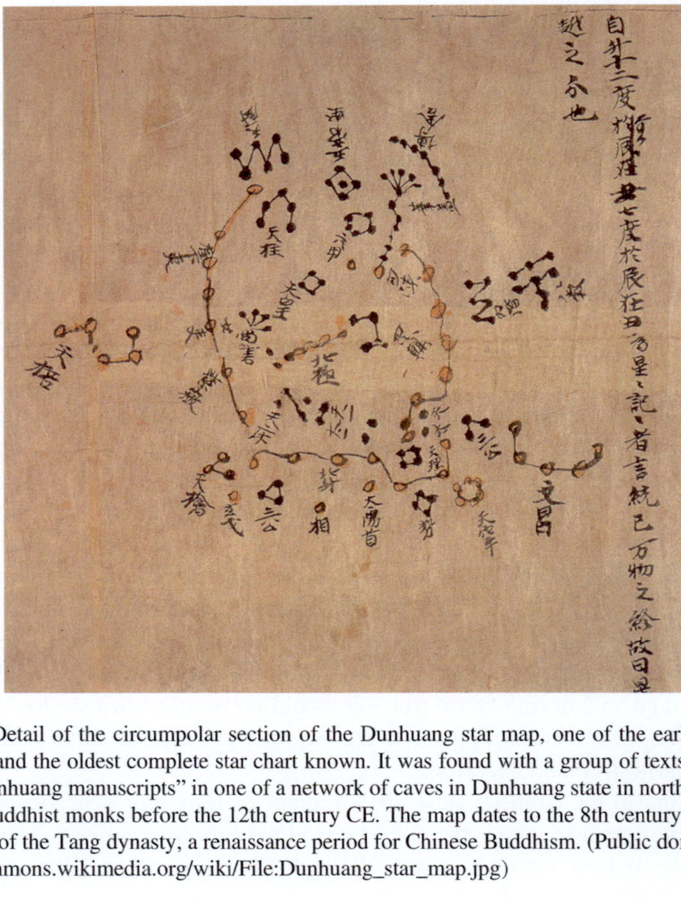

Fig. 4.4 Detail of the circumpolar section of the Dunhuang star map, one of the earliest known star maps and the oldest complete star chart known. It was found with a group of texts referred to as the "Dunhuang manuscripts" in one of a network of caves in Dunhuang state in northwest China used by Buddhist monks before the 12th century CE. The map dates to the 8th century CE, during the period of the Tang dynasty, a renaissance period for Chinese Buddhism. (Public domain photo, https://commons.wikimedia.org/wiki/File:Dunhuang_star_map.jpg)

BCE was likely Beta Ursae Minoris, in what we know as the "Little Dipper" (Ursa Minor) constellation. The stars surrounding the pole star, or the celestial image of the emperor, were referred to as the walls of the purple forbidden palace, after which the palace built in Beijing, also called the "Purple Forbidden Palace" was designed in the early 15th century. Even today in the 21st century, anyone who visits the Forbidden City is awed by its majestic grandeur, and most of all by its sheer enormity. It seems to continue forever, in all directions. Its high walls give the impression of an enclosed courtyard complex, but continuing through it one discovers courtyard after courtyard, complex after complex. It gives one a sense of the vast scale of the heavens, and certainly a suitable structure for creating in its visitors the sense that one is indeed in the center of a celestial empire. And well before the construction of the Forbidden City, Chinese rulers built cities intended to mirror the structure of the circumpolar region of the sky. The city of Chang'an, for example, constructed as a capital during the Western Han, followed this pattern.[7]

While there were many different conceptions of the structure and organization of the cosmos, three of these theories in particular enjoyed more influence than the others, and in various periods were adopted by important astronomers, thinkers, or state organizations. These three systems were the *gai tian* ("Covering of Heaven" or "Heavenly Dome") theory, the *hun tian* ("Heavenly/Celestial Sphere") theory, and the *xuan ye* ("Infinite Space") theory. Interestingly, just as is the case with the equatorial nature of our astronomy, we find another ancestor of our contemporary understanding of the universe and our cosmology here, in the *xuan ye* system, which eventually gained ascendancy over the other two cosmological systems.[8]

The *gai tian* system seems least consistent with the equatorial nature of most of the history of Chinese astronomy, and indeed enjoyed the least influence of the three theories (although there were those who endorsed it). This view took the earth to be a flat plane covered by the dome of the sky, which moved consistently with what we observe in the sky. The seeming virtue of this cosmological picture and other views like it is that it seems the most consistent with what we actually observe when we view the skies from our vantagepoint on the earth. The earth appears to us "flat"—that is, roughly in a plane, and the sky appears to us as a dome, surrounding us on all sides as well as overhead. Why a dome, we might ask, rather than a cube? There are a couple of reasons. The main reason is that the motions of objects in the sky—the sun and moon, planets and stars—are in arcs rather than in lines. If the sky were a cube rather than a dome, we should expect the motions of objects "on" this cube to be linear rather than in arcs. But what we in fact observe is that the sun, for example, rises from the horizon and follows a curved arc until it sets in the horizon on the opposite (western) side. Since the sky surrounds us on all sides and above,

[7]Nancy Steinhardt discusses this and a number of other imperial cities, including astronomical patterns in construction, in (Steinhardt 1990).

[8]These systems are most extensively discussed in relation to one another in (Needham and Ling 1959).

the best way to make sense of this along with the fact of the curved motion of celestial objects is that the sky is a dome rather than a shape with edges.

The downside of the *gai tian* theory, like any theory that maintains that the sky is a dome, is how we make sense of the peculiar motion of the heavens, and also how we make sense of the fact that objects that disappear on one side of the sky reappear on the other—whether the sun, moon, stars, or planets? First—in most latitudes, such as that of all of China, and specifically of the region of China in the ancient period, which surrounded the Yellow River in the northern part of the modern-day nation of China, movement of celestial objects will be around a pole that is not located anywhere near the center of the sky. Only at the north pole and the south pole, where no one lives, will there be a sky in which the pole of celestial motion is at the zenith, and the celestial objects turn around this point. If there were a civilization that could survive at either of the poles, it would be natural for them to adopt a view like the *gai tian* view, as it would indeed appear that the motion of the celestial objects were consistent with being on or within a dome. Elsewhere, the pole of motion will be removed from the zenith. For the reasons explained above, at the center of Beijing, for example, the pole will be 39°, 54 min, and 50 s elevated from the horizon in the north. Thus, the motion of the skies will be offset to this point, rotating around it, rather than the zenith, and objects such as the stars will disappear below the horizon and rise from the other side of the horizon in later nights. This motion is not suggestive of the movement of objects on or within a dome. In addition, what do we say happens to the sun when it descends into the west and goes beneath the dome of the heavens? Is it annihilated and reborn at the beginning of the next day? And if not, how does it get from its position of disappearance in the west to its position of reappearance in the east the next morning? Does it travel around the side of the periphery of the dome? Does it circle underneath the dome? But if it does the latter, this seems to suggest that the sky is not a dome after all, but a sphere. That is, if the sun can move in an arc *below* the dome of the sky in order to get back around to the east to rise again in the morning, than its motion on the opposite side of the dome must be the same in kind to its motion across the dome as we witness it. Thus there must be *another* dome opposite our own, an anti-dome, if you will, but if you connect two domes to one another at their bases, you create a sphere. And it is just such a cosmos that is offered by the *hun tian* theory.

According to the *hun tian* view, the cosmos is comprised of the sphere around the earth through which the celestial objects move. This position dissolves the problems with the *kai tian* view, while still explaining why we see the sky the way we do. We observe only one half of the celestial sphere, and the opposite half is invisible to us, because it would only appear to observers on the opposite side of the earth. The *gai tian* view solves the problem of apparent motion on a dome, but it entails a much different view about the Earth itself than what we find in the *hun tian* theory. If the cosmos is a sphere surrounding the earth, perhaps the Earth is also a sphere at the center of this celestial sphere, such that there might be observers on the other side of it, watching the motions of the sun and other celestial objects during the times they are invisible to us, on our side of the Earth sphere. The ancient Greek astronomers had a conception of the cosmos very similar to the *hun tian* view throughout the ancient period, and this conception remained dominant in Europe all the way until the modern period, when the Copernican revolution finally overturned it. From Aristotle's notion of the "heavenly spheres" to Ptolemy's formulation and even Tycho Brahe's attempt to save the view of the spheres by making Ptolemaic views consistent with observations that seemed to undermine it, the West accepted a view roughly equivalent to that of the *gai tian* cosmology, and along with it a number of false views about the heavens that led to the failure to recognize and understand the nature of certain celestial phenomena like supernovae and comets until a relatively late period—two phenomena the Chinese understood as non-meterological, and (in the case of supernovae) even extra-solar system events.

The most influential cosmological view for the history of Chinese astronomy, and one roughly similar to the contemporary view of the cosmos, was the *xuan ye* view. It was the adoption of this view that made it possible to understand the true nature of very important celestial events, and to understand the full extent of the cosmos, at least to a greater degree than would have been possible on the other two cosmological views. According to *xuan ye*, the cosmos is made up of vast, indeed *infinite*, empty space. The sun, moon, planets, stars, and all other observed events in the heavens take place in this immeasurable space. There is neither a sphere nor a dome around the Earth, rather the earth sits, like the other celestial bodies, within an infinite amount of space, and other objects are distant from the Earth by varying degrees. The stars are far away, and this is why they are not observed to move relative to one another, while objects like the planets, sun, and moon are closer. Ultimately, all of this exists within the shapeless and infinite container of space that is the cosmos.

This may, and should, sound very familiar to you, as this is very close to the conception of the cosmos astronomers and most of the rest of us hold today. When we talk about the cosmos outside of the Earth, we use the term 'space'. Space is not a dome, nor is it a sphere, but it is an immense area in which there is matter, that clumps into stars, planets, and everything else that exists. Perhaps the only difference between the contemporary view and the *xuan ye* view is that according to *xuan ye*, space is infinite, and thus not bounded. On the contemporary view, space is not bounded, but it *is* finite. There is not an infinite amount of space. Even though the amount of space is enormously (indeed, astronomically) large, it does have a limit. It does not go on forever. Still, the *xuan ye* cosmology is closer to that of the modern world than most premodern cosmological systems, including that of the astronomically adept Maya. The *xuan ye* cosmology gained ascendancy in China relatively early on.

The existence of astronomical tools such as the armillary sphere from early on in Chinese history seems to show that at least the astronomical class never took very seriously the idea of *gai tian* that the sky is a dome over the earth. The armillary sphere accurately depicted the poles and the equator, the ecliptic and the meridian. Anyone who regularly followed the motions of the heavens would recognize their inconsistency with the view that the sky is a dome, and spherical motion would make the most sense of the data. The early Chinese astronomers knew the world was a sphere, throughout the entire recorded history of the region. While the *gai tian* view may have been accepted by some uninformed non-astronomers, the intellectual classes and astronomers never took it seriously. The *hun tian* theory, on the other hand, was endorsed to some extent by the greatest astronomers, such as Zhang Heng, who, although he accepted the theory as consistent with observation, also recognized its deficiency as a complete theory of the cosmos. Zhang seems to have accepted both the *hun tian* and the *xuan ye* views. How can this be consistent, we might ask? Well—one might adopt or endorse certain positions for their practical value while recognizing that they are not representative of the ultimate truth. The armillary sphere was a powerful instrument for observation, and the *hun tian* theory made most sense of its motion and the observations consistent with it. One might provisionally adopt such a picture of the cosmos, with the knowledge that it is not exactly or not technically correct, or does not give us a complete picture of the truth. Physicists in the contemporary world do much the same thing. We have known for quite some time (since the early part of the 20th century) that Newton's Laws are not universally applicable. Classical Mechanics break down when dealing with the very large, the very small, and the very fast. Einstein demonstrated this in his theories of relativity, and observational data proved him right. Yet classical mechanics work perfectly well when working with objects, forces, etc., that we encounter on earth. Thus, classical mechanics is still taught in physics courses not as a historical relic, but as an applicable and acceptable for understanding the world.

It was because of the *xuan ye* view that the Chinese astronomers were able to recognize comets as something outside of the atmosphere, in the vast infinite space of the cosmos. They did not have any "perfection of the spheres" to protect, as they may have if they'd adopted the *hun tian* view. Not only did the Chinese astronomers understand much of the nature of comets (although not with the in some respects fuller modern understanding), but they also held them to be one of the strongest portents, rare events that could mean either victory or doom for a fledgling dynasty such as the *Xin*.

Court scholars were relatively common in ancient China even before the institution of empire, in the Spring and Autumn and (especially) Warring States period. The Warring States philosopher Han Feizi, for example, was a minister to the ruler of the state of Qin—the same state that eventually conquered the other warring states of the region and established the Qin Dynasty, which, although it was just about as short-lived as Wang Mang's Xin dynasty, at least has the pride of place of being the *first* imperial dynasty, and so has a central place in historical memory far beyond that of the Xin (even many non-Chinese have heard of the Emperor of Qin and the "unification" of China. How many have heard of Wang Mang?). Exploiting this feature of his newfound dynasty, the ruler of Qin after conquering the warring states gave himself the title of *Qin shi huangdi* (First Grand Emperor of Qin). Not a very humble act—the term *huangdi* really goes beyond our term ' emperor'- it is something that even has religious connotations. This was akin to the ruler of Qin giving himself the name "The First God of Qin".

In the Han dynasty, one of the major intellectual figures, and another court scholar, was Dong Zhongshu, with whose name is associated a collection of ideas of the early Han Dynasty, *Chunqiu fanlu* ("Luxuriant Dew of the Spring and Autumn"), a collection that includes statements of the new Confucianism that became popular in the Han, but also includes some explanation and theorization concerning the peculiar metaphysical and cosmological view prevalent in early China that allowed astronomers to make connections between celestial events and the fate, good or bad, of rulers and their dynasties. The ways of Heaven could not always be known by humans, even though the two were inextricably linked, as philosophers such as Dong Zhongshu argued.

"Heaven and Humanity form one unity," Dong had written early in the Western Han. Through the various actions of nature, we could observe the quality and state of the human world. When the body is ill, the mind inevitably suffers as well, we all know. We could think of humanity and the rest of nature in the same way, Dong and his contemporary thinkers claimed. This extends to other features of nature as well. Just as weakness of my body indicates lack of *qi* or fatigue, external signs in nature indicate certain states of the human community, as inextricably linked to this nature, as one part of nature. Because of this, in order to understand the human realm, including the current state, as well as the past and perhaps most importantly future states (which rulers, like businessmen, would love to know) of persons, one could study the state of nature, and read in its changes and motions the changes and motions of the human world. Catastrophic events in nature signaled catastrophic events in the human realm, and likewise for auspicious events. Whoever understood nature held the power to divining the future. And no aspect of nature was more significant than the sky. This is the reason the Chinese were the premier astronomers of the entire globe for most of human history. Understanding the heavens was vital to survival. Given the unity (*he yi*) between humans and nature, also often understood in terms of resonance (*gan ying*) between the two, such that actions in one affect another, just like plucking a guitar string creates a vibration, which produces a sound, understanding and being able to forecast the effects in the human world of celestial events became of crucial importance.

Chinese Astronomical Systems and Tools

Those whose job it was to search the sky, discover and document the events happening there, as well as interpret them, were a professional class of court astronomers, whose learning was focused and developed. They constructed sophisticated tools for observing, developed complex and accurate coordinate systems, and devoted their entire energies to the study of the sky. Astronomy throughout the history of China was a devoted professional activity. It was not done by amateurs who spent most of their time in other work, whether working fields or governing the state. Perhaps this is another legacy of the Chinese astronomers in the modern world. An astronomer was (with exceptions) someone devoted specifically to astronomy. They would generally come from the literati class, so would be educated in all the classic necessary for inclusion into that group, and, unlike the merchant or farmer classes, they would not labor in this way for their living. Rather, they would be supported by the government (imperial or regional), employed to focus on astronomy.

One of the unique features of Chinese astronomy is its coordinate system based on the celestial pole. As mentioned above, the pole plays an enormous role in Chinese astronomy, and understanding it and the movements around it are critical for an understanding of Chinese astronomy. One of the reasons for the supreme importance of the pole in Chinese astronomical thought is its identification with the ruler of the world. The emperor himself is represented by the pole star, around which all other stars rotate. It reminds one of Aristotle's "unmoved mover"—the being that is the cause of all motion, but itself does not move, and is first cause of all things. Or an image closer to home, that of the *Analects* of Confucius, in which Confucius describes the ideal ruler as one who is like the pole star, which sits in its place without moving, and around which all the other stars move and pay tribute. The "resonance" between the sky and humanity can be understood in some ways as offering us a *map* of the human world—the sky shows us in this way the ideal hierarchy, the positions of everyone in the world relative to one another, which it is our job to duplicate.

The coordinate system based on the pole used by the Chinese divided up the sky into 28 sections, and defined these sections by lines running from the pole to and through the equator. The positions of these sections and lines were determined by circumpolar stars and constellations—so that one could easily determine the position of a celestial event or object in a way somewhat similar to what we today call "star hopping". Star hopping is to find an easy to locate constellation or star, and use it to find some less apparent object. One may find the Big Dipper, for example, near the pole, and find an object by looking across the sky in an invisible line continuing from the handle of the dipper, in order to find a less apparent object roughly along such an imaginary line. Such a system of lines is just what the Chinese astronomers used as the basis for a coordinate system. Their use of these lines was much more systematic and particularized than our idea of star hopping. They set a fixed number of lines attached to particular stars, and defined a region of the sky on the basis of this star, and its connection to other stars nearer the equator. These sections dividing up the equator were called *xiu*. These *xiu* could be determined by stars near the equator, but one thing that we notice when we look into the night sky is that bright stars that might be used as guideposts don't always fall on or near the celestial equator. Thus, positions of certain other stars on a line connected to a particular circumpolar star were used to determine a *xiu*. There were 28 of these *xiu* determined by the Chinese astronomers, and they were seen as *houses* of the astronomical objects that could move between them, namely, the sun, moon, and planets. *Xiu* is sometimes translated as 'mansion' in this regard. The *xiu* in particular may have been based on lunar motion, as we see that one full orbit of the moon around the Earth, and thus the time from new moon to new moon as witnessed by us as well, is about 29 days (to be exact, 29.53 days). At the same time, its sidereal period (that is, its movement around and back to the same position in the sky relative to the stars) is somewhat shorter, about 27 days (27.33 days). The number of *xiu* could thus be an attempt to indicate the daily motion of the moon, and for this reason the system has often been referred to as that of the "lunar mansions."

As the moon on its way around the earth occupies one of the *xiu* at a time and the sun also occupies a *xiu*, although its progression through the 28 *xiu* takes a full year, rather than the lunar month. Just as in the zodiacal system of the Greeks and others, the sun's position at any given time can be given based on the *xiu* that it presently inhabits. It is a relatively easy task to determine the *xiu* of the sun even though we of course cannot see the stars when the sun is visible because its light illuminates our atmosphere and blots out the stars. Determining the *xiu* of the moon is simple, as we can see the stars surrounding it, and from just a quick glance (with knowledge of the 28 *xiu*, of course, we can place its location. With the sun, a different technique is necessary. The sun passes through its highest point in the sky, regardless of what time of the year it is, at noon. Although the elevation of this point will change, being highest at the summer solstice and lowest at the winter solstice (outside the tropics at least), it will always be the case that noon marks the highest ascent of the sun in any given day. Since this is the case, the sun will be one half of the way through its daily motion around the Earth (actually the Earth's spin on its axis, of course) 12 h later, at midnight, halfway through the 24 h period of noon to noon. Thus, in order to determine which *xiu* the sun currently occupies, all we need to know is which *xiu* presently crosses the meridian at exactly midnight, and in addition we need to know which *xiu* are directly opposite in the sky to that *xiu*. Thereby we can place the sun.

This method could be generalized to be even easier. One need not stay up until midnight on any given night to figure out what *xiu* the sun occupies. First, it would be relatively easy once we have observed the sun's motions for a year to determine which *xiu* it will occupy during any given time of the year, as this motion will remain static. The tilt of the earth remains (roughly) the same, as does the orbital motion and period of the Earth in its movement around the sun. Just as in the zodiacal system of the Greeks, the different parts of the year could be determined to be associated with the occupation of the sun of different *xiu*. Interestingly enough, the Chinese were never nearly as concerned with the *sun's* motions through the *xiu* as were the Greeks.

The *xiu* were one component of the Chinese coordinate system, marking an object's position along the axis of the equator. Since the *xiu* were determinative of regions of space based on divisions of the equator, they could only be used to determine one half of the coordinates of any given object—its latitude on the celestial sphere, equivalent to the contemporary measure of right ascension, left and right along the equator. One easy way to think of this is to think of it in terms of a Cartesian coordinate system, on a flat surface—one of the kind you've likely seen in school. In such a coordinate system any point is located by an x and y position. To determine the location of any given point, we have to determine where it is along each axis. Every point, including the center point and all points along either of the axes, have both x and y coordinates. Thus, the center point has the coordinates x = 0, y = 0, while a point along the x axis may have coordinates like x = 4, y = 0. If we take a Cartesian coordinate system and turn it inside out so that it stretches into a sphere with us inside it at the center, where the x axis is the equator and the y axis is the meridian, we then have an equatorial coordinate system, like that used by the Chinese and that used today. A position was determined along the equator, in the Chinese system, by the distance of an object from a *xiu* determinative star, in degrees. So, for example, the equatorial component of the coordinates of a particular object may be 3° west of a particular *xiu*. For declination, the Chinese used the measure of distance from the northern celestial pole. This provided the y axis coordinate of elevation from the equator. In this way, using the *xiu* and northern celestial pole, the position of any object in the sky could be determined.

The *xiu* determinative stars and constellations could give an indication of the position of the moon, sun, and planets. A key difference between the Greek system and that of the Chinese is that, due to the equatorial nature of Chinese astronomy, the important constellations were equatorial, rather than the ecliptic constellations for the Greeks. The main reason that the Greeks focused on ecliptic constellations and the ecliptic in general is that their primary concern in astronomy grew out of a consideration of the motion of the sun and the planets. The moon does not travel on the ecliptic, as its orbit is elevated to the ecliptic. A concern with the motion of the moon in part led to the *xiu* system. Although the position of the sun can also be determined in terms of *xiu*, this will not be as exact as positioning the sun via constellations along the ecliptic, as the sun "moves" through the ecliptic (just as lunar positioning via ecliptic-based constellations will be equally inexact). In the Greek system, the sun at any given time of the year occupies some constellation along the ecliptic, in that it covers those background stars. Thus, in July or August the sun is *within* the constellation of Leo—or at least, if we could turn down the sun's light so we could see the stars, we would see the sun against the background of the stars of Leo. In the *xiu* system for determining the sun, it will often happen that the sun is not *exactly* within the constellation determinative of the *xiu*, but is in the segment of the sky (one of 28) determined by some given *xiu*.

Another celestial object that received attention in Chinese astronomy was the planet Jupiter. Just as the planet Venus had a great significance for many cultures of the Americas, Jupiter was an important planet for the Chinese. There are a couple of major differences between Jupiter and the other naked-eye visible planets that seem to naturally make it an object of interest for people with a lengthy astronomical history. First, since its orbit is much longer than that of the inner planets, being much farther away from the sun, it is both slower, and has longer to travel, which makes Jupiter's movement through its orbit around the sun much longer than ours or those of any of the inner planets. One year for Jupiter (one full orbit around the sun) takes about 12 Earth years. All of the planets outside of the Earth's orbit, will of course take longer to orbit the sun than does the Earth, while those within our orbit, specifically Venus and Mercury, will have shorter orbits and travel faster, thus having shorter years. Secondly—Jupiter is the brightest (by far) object at its distance, far outshining the only other outer planet that we can see with the naked-eye,

Saturn. There are of course a few reasons for this. Jupiter is closer to us than is Saturn, and Jupiter is also far larger than any other solar system object other than the sun itself. Jupiter always appears as a fairly bright object in the sky, but shines even more brilliantly when it is closest to us. Because Jupiter is outside of our orbit, it does not go through phases (at least from our standpoint—if we were on Saturn or Neptune, we would see Jupiter undergoing phases).

The Chinese "zodiac", in yearly cycles, is based on the movement of Jupiter through its orbit. Each year, Jupiter will traverse 1/12th of its orbit, and thus the planet can serve as a good indicator of a period of time close to our notion of the "decade". Think about our own relationship with decades. We fix periods in our lives and our culture in terms of decades. We associate certain cultural and intellectual trends, historical events, and personages with decades—the 1960s, say, or the 1980s. Generally, decades are more meaningful to the individual within or close to his or her own lifetime—such that unless one is a student of history, the decade of the 1840s or the 1610s will not signify very much for them. On the other hand, there is hardly anyone alive for whom the 1960s, 70s, and 80s have no connotations. The motion of Jupiter is one natural way of fixing such periods, and an interesting astronomical alternative to our measure of decades. The zodiacal animals associated with the Chinese New Year are connected to this motion of Jupiter. Since Jupiter moves through 1/12th of its orbit every year, it will also occupy a different *xiu* every year (it will go through *two xiu* per year), and occupy a different constellation at the New Year every winter. The New Year is defined in China by the position of the moon—the first full moon after the winter solstice marks the New Year. The New Year celebration, as many know, has through Chinese history become the central major holiday in Chinese culture (imagine Christmas, Thanksgiving, and New Year's Eve wrapped into one).

Jupiter will occupy a different constellation each New Year, and the year is determined by the constellation in which Jupiter resides at that New Year's day. Thus, on New Year's Day in the year I was born, 1978, the sun resided in the Horse constellation, and so I was born under the sign of the Horse. One's Chinese birth sign, as well as the year, is determined by this positioning. 1978 was the year of the Horse, and 2014 was again the year of the Horse, the year of my birth sign, as it has been 3 other times since I've been born (1978, 1990 and 2002).

So we see that the positions of the moon, sun, and planets were important and could all be determined by the polar and equatorial positioning system created by Chinese astronomers. Another feature of celestial positioning important for the Chinese astronomers was the placement of celestial objects with respect to other important objects. With the appearance of a comet, supernova, or the occultation of a planet, for example, it was significant where in the sky this happened, because its occurrence in a particular region or the environs of a particular star or group of stars would generally have some kind of significance in the resonance between nature and humanity. If the stars represented a mirror of the human world, then a fantastic celestial event happening in some particular region of the sky signaled some major event (either a boon or a catastrophe) to happen in the equivalent part of the human world. A bright celestial event like a supernova or comet appearing near the pole star, the abode of the emperor, would have been particularly ominous—although one imagines that sufficiently creative court astronomers could have massaged the interpretation so as to explain how the event augured a coming age of even greater glory for the emperor and his dynasty.

How did the early Chinese understand particular events such as the great comet of 178, in context of the continuing reigns of the celestially identified rulers at the head of the civilized human world? The major events that moved people to a consideration of the meanings of celestial motions were, throughout Chinese history, in large part the same as they are today—comets, supernovae, and eclipses of the sun and moon. Chinese astronomers in the second century CE had quite a number of sophisticated astronomical tools to aid them in their investigation of the heavens for signs related to the human world. Some of the most ingenious and effective observational tools in human history were invented and used by the Chinese, and as I discussed above in the case of the equatorial armillary sphere, there are a number of them that are versions of instruments still in use by astronomers today.

To position in the sky objects like comets, one had to have a sense of the time as well as the position against the background sky, as this would tell you where this object was with respect to the sun—information that could be useful especially for comets. Of course, one could always determine this by knowing which *xiu* the sun occupied, but in order to accurately determine this, accurate timekeeping was also needed, to determine meridian passage of a *xiu* determinant star at midnight. Just as astronomy required (and still requires) time, so also astronomy was required to *create* time. The connection between astronomy and time is as old as humanity. In most cultures worldwide and throughout history, the basic unit of time, the day, was measured in terms of the sun's passage, a complete spin of the Earth on its axis. Even today, when we use the second as the basic unit of time, and define this based on the atomic clock standard, this can also be thought of as deriving from the astronomical measure of the day. There were seconds long before there was the atomic clock, and our contemporary seconds are a formalization of those earlier seconds. Seconds were ultimately just smaller components of a day, allowing us to be more precise with our placements of time. A day was broken into 24 h, those hours into 60 min, and those 60 min into 60 s. This system was based on the sun, and is the system we still use, even though we have found far more efficient and regular ways to determine these measures of time than the motion of the sun around the earth.

The Chinese astronomers had a number of different ways of determining time, some based on the sun's motion and others more efficient. The most basic way of determining time, at least during the daytime, is to use some version of a sundial, which operates via the function of a gnomon. I mentioned briefly the gnomon in the first chapter above, describing its use at North American sites like Cahokia and SunWatch. The early Chinese also used gnomons, but due to the different character of their astronomy, they used them in a different and in some ways more efficient way. The most basic gnomon, of course, is simply a pole or stick in the ground, which casts its shadow in a particular direction based on where the sun sits in the sky at any given moment. Such a gnomon works fine as a way of determining solstices, and can even be serviceable as a time-keeping instrument, although an inexact one, as the lengths of the shadow cast by the gnomon and thus the space carved out in each hour of the day by it will be different.

One way to eliminate this issue of variable shadow lengths and movement times is to tilt the gnomon so that it is not facing straight up, but points directly to the celestial pole. The Chinese, given the polar and equatorial nature of their astronomy, found this extremely natural, and were the first to develop the equatorial sundial. In such an instrument, the gnomon is pointed at the pole, and an additional equatorial plane is needed on which the gnomon can cast its shadow. This plane, of course, will run perpendicular to the gnomon, and will thus be continuous with the plane of the celestial equator. We see all kinds of materials used for this, and various different ways of creating the equatorial plane and gnomon. One way of doing this is to create a ramp that angles at the tilt of the equator, then fix a pole or another angled ramp whose edge points at the pole, exactly perpendicular with the plane of the ramp. We will see in the next chapter perhaps the most magnificent example of such a tool in existence, created by the Maharaja Jai Singh II of Jaipur (in current day Rajasthan, India).

While the Chinese developed such sophisticated sundials, there is another problem inherent in the sundial or any other tool for telling time based on the motion of the sun. The motion of the sun through the sky, that is, is not constant throughout the year. The explanation for this has to do with the *eccentricity* of the Earth's orbit. Unlike what was posited by Copernicus, the Earth does not travel around the sun on a circular orbit, but instead follows a path much closer to an ellipse (or oval). At the same time, as Johannes Kepler demonstrated that a planetary body orbiting the sun covers an equal amount of area (thought of in terms of the pie slice of the entire plane of orbit rather than the distance traveled) in equal time. Because of the eccentricity of the orbit, this means that during certain periods of the year the Earth will move faster, covering more distance than an equal period of time elsewhere in the year. Because of this, the discrepancy between apparent solar time, that is, the time the sun takes to go around the earth, and mean solar time, or the average solar time distributed throughout the year, such as the time kept by a standardized clock that ran the same all year long, will diverge. Because the Earth will be moving fastest when it is closest to the sun, and more slowly when further away, during the time of the year the Earth is closest to the sun in its elliptical orbit, the apparent solar time, or movement of the shadow of a gnomon through the hour lines of the sundial, will be faster than the mean solar time. At the closest approach of Earth to the sun, which happens in January and February, the apparent solar time will be much slower than mean solar time, with the difference amounting to the startlingly high number of 15 min! Thus, using a sundial in February will give one a time 14 min earlier than the actual time (thought of as mean solar time). In November, on the other hand, the sundial will give a reading of more than 16 min faster than the actual time. The eccentricity of Earth's orbit is not the only factor contributing to the problem. Because of the obliquity of the ecliptic—that is, the fact that the plane of the ecliptic, which is the plane on which the Earth and the other planets orbit, is not the direction the planet's equator faces, due to the Earth's being tilted on its axis, the position of the sun for an observer on any given spot on the Earth will appear to move throughout the year. This is what is responsible for the seasons, for the movement of the sun from low in the sky at the winter solstice to high in the sky at the summer solstice. It also explains why seasonal changes are not as pronounced in the tropics, where the movement of the sun due to the obliquity of the ecliptic does not take the sun very far from the center of the tropical skies.

These two effects—obliquity of the ecliptic and the eccentricity of our orbit, combine together to create the divergence between apparent and mean solar time. This is a problem of any sundial. To fix the problem, we can use what is called the *equation of time*. If we know the effects on the difference between the two solar times due to the two relevant effects, we can calculate the difference, and simply add or subtract it to the observed apparent solar time we read on the sundial at any given time of the year. The equation of time can be represented by a simple graph, and this graph can be (and was) often contained alongside sundials that were meant to be functional timepieces.

The Chinese astronomers, as well as normal people who wanted to read the time, used a wide variety of different sundial based timepieces throughout history. These devices ranged from the enormous fixed sundials that one might find adorning an imperial court, objects that were as much lavish decoration as they were functional astronomical tools, to the portable devices that could be held in the palm of a hand and positioned in the right way so as to get a fairly accurate determination of the time. An additional feature of the sundial or gnomon that can be taken advantage of is the added precision one can gain with such a device by making it larger. The larger the device, the more movement of shadow will take place over a larger area, which allows observers to make finer distinctions in position of the shadow, as they become able to break each component into finer segments. The reason this happens is because for any sundial the shadow will take the same amount of time to travel through its motions, as the sun's motion through the sky is constant, and the same for any sundial. A larger gnomon will cast a larger shadow that moves through more area, and thus the shadow will move across the same distance faster than the same distance is covered on a smaller sundial (distance in terms of linear distance, not in terms of percentage of percentage of daily motion, which will be the same). In addition, one will be able to mark positions and transitions with greater detail because of the greater space traveled by the shadow of the larger gnomon. For example, if a shadow on a small handheld sundial passes through 1/12th of its arc (let's say on one of the equinox days), this constitutes one hour of time. If the distance from end to end of the sundial is only one foot, the distance the shadow moves over this hour will be less than an inch. It will be hard to gain much more detail about the time than what hour it is at any given time, as neither the detail of the shadow nor the power of our eyes is great enough to see 60ths of an inch (minutes), let along 3600ths of an inch (seconds). If we increase the size of our sundial such that the distance passed through is 500 feet, then in 1 h the shadow will move through about 40 feet! Plenty of distance for us to be able to resolve minutes (8 in.) and estimate seconds. Thus, the larger sundial is a much more accurate timepiece than the smaller one. However this is not always the case. Based on size alone, the shadow cast will move more quickly over a larger area, but it will also be the case that the shadow of the gnomon's edge will become more diffuse. Early Chinese astronomers came up with an ingenious solution to this problem in their creation of the "shadow definer" (*ying fu*), which uses the pinhole lensing effect to create a sharper image of the gnomon shadow edge. This principle was used, to amazing effect, by the Chinese astronomer Guo Shoujing, who erected the enormous gnomon tower of Zhou Gong in 1276 CE, a structure that was exclusively dedicated to measurement of the length of the sun's shadow. In India, the idea of large astronomical instruments such as gnomons and sundials was brought to its natural end with the colossal instruments of Jai Singh II, which we will look at in the next chapter.

A number of other tools could be used to read time at different times of the day, including at night when the sun was not available. In the Middle East and the West, the well known object for finding time during the day or the night was the *astrolabe*, an object that became perhaps the most well known and widely used astronomical tool in the history of the West. The astrolabe worked by in a sense allowing the user to use the locations of stars, given certain times of the year, to determine the time. After all, the shadow cast by the sun on a sundial is based on the position of the sun in the sky, which we cannot (or at least should not) observe directly. With the stars at night, we can directly observe, and we can take any given star or constellation as representing the time-determinative object, as long as we know at what time that object crosses the meridian. For visible objects, we can use objects that we know cross the meridian at midnight as determinative of current time. When such an object is halfway between the horizon and the meridian, we can determine that it is 9 pm, for example.

Two other tools of note in China for determining positions were the sighting tube (often made a part of an armillary sphere) and the *xuanji*, a polar alignment tool (Fig. 4.5).

The *xuanji* was a small device one could hold to the north polar region, that would help one to determine the lines with which the various *xiu* sections of the sky were divided. One can see how such an object would be of help when trying to determine the *xiu* position of a particularly high declination object, too distant from the equator to be easily placed into a particular *xiu*. With such an object, one could determine the proper sight lines or *xiu* lines based on the pole, and thus place such objects as northerly comets and other objects and events. There would have been much occasion to do this, as we hear accounts in the Chinese records of comets that appear at or near the pole—no doubt significant political events, given the association of the pole with the emperor!

The sighting tube was another ingenious device that allowed for viewing of particular regions of the sky, and particular objects such as stars and comets. One of the most amazing features of the sighting tube is that it allowed for greater resolution of celestial objects—a kind of ancient world telescope! While the sighting tube did not increase the magnification of a particular object—the disc of a planet

Fig. 4.5 The *xuanji*, a small jade disk with a series of "teeth" along the outside ring, was likely used as a tool for locating the pole. In ancient times, our familiar "North Star" of Polaris did not inhabit the pole, and for much of human history there would have been no star at the pole. Even today, Polaris does not lie directly on the pole, but is close enough that we discern no difference with the naked eye, and it can be used to fix the pole

would not appear larger, for example, it did increase the amount of light that could be seen in any given region of the sky. In general, in dark skies of the kind that would have been available in ancient China, the human eye can resolve objects down to about magnitude 6, for those with excellent eyes (a good test of eyesight is to look at Epsilon Lyrae in the late spring and summer sky—those with very good eyesight will be able to do something I personally am unable to—resolve the object as a binary star). The sighting tube can increase the magnitude resolvable by as much as two points of magnitude! Thus, someone who can resolve objects down to magnitude 6 will be able, using a sighting tube, to resolve them down to magnitude 8. How does this work?

The key principle at work in a sighting tube is essentially the same as that at work when we create domes for telescopes to keep out surrounding light, or the reason that we are able to resolve more stars at greater levels of dimness and higher magnitudes when it is darker out and there are fewer lights. The more light there is entering our eye, the less we will be able to see from dimmer objects that are drowned out by this light. This is the very reason we are unable to see stars when the sun is up. The stars are still there, of course, but the brightness of the sun lights up our atmosphere, as this immense light is scattered, and creates a brightness that simply drowns out the stars. It is not only during the daytime that this happens. Walk into the downtown of any major city on a clear night, and look up into the sky. You will not see very much at all, beside perhaps a couple of the very brightest objects in the sky, such as Venus and the two or three brightest stars. The reason for this is the same as the reason we cannot see the stars during the day—the surrounding light is far brighter and simply washes out the dimmer celestial objects. Even during the day, and in the middle of the city, the light from the furthest stars is reaching us, and the information from this light is there to be read if we have the proper tools—but our eyes are not the proper tools. There are many objects in the universe whose light reaches earth but which are far too dim—that is, send us far too little light, to be seen by the naked eye. If this were not the case, we would not be able to resolve these distant objects in our powerful telescopes. Telescopes can only receive light that is *there*—and make this dim light resolvable to our eyes.

The sighting tube is created by putting a pinhole sighting device at the end of a tube that can be fixed to an armillary sphere or some other object that can be moved around to focus on different spots in the heavens. The sighting tube can be handheld as well, but as anyone who has tried to view the heavens through binoculars while holding them can attest, this is easier said than done. One's hands are a much shakier mount for a scope than any inanimate object. What the pinhole sight at the end of the tube does is to block out the light surrounding the object being viewed, so that one can view an object more directly without the intervening light from surrounding stars and other objects. This, in effect, increases the eye's ability to resolve the object in the sights, because interfering light is blocked out, and the light received from the object can be more directly observed. Thus, the sighting tube was, before the invention of the telescope, the instrument with which one could resolve the dimmest possible objects in the night sky. And it is their use of the sighting tube, in part, that allowed the Chinese to discover and document comets and other celestial objects that were missed elsewhere in the world. In many parts of the world, people would only have seen the major comets, bright and impossible to miss—comets like Halley's, like Caesar's Comet, and the great comet of 178. The Chinese astronomers documented many more—tiny, distant comets, almost too small and dim for the eye to see. The aim of the skywatchers of the Celestial Empire was to see all, and record all.

The comets, droughts, and other events that spelled doom for Wang Mang's *Xin* dynasty were not completely uncommon in Chinese history. Not only was it important what happened, but also *when* it happened. The same events that took place during Wang Mang's reign, if spread into one of the two Han dynasties, may have made no difference at all. But certainly these events happening in the closing years of the 2nd century CE would have been noticed. Like Wang Mang's *Xin* before it, the Later Han was dissolving, and as always, the answer could be found in the sky.

In the last days of summer in 178 CE they, and many other people around the world, saw the incredible comet that stretched across the sky, its massive reddish dust tail stretching through several degrees of the sky. Within the Celestial Empire, the second Han dynasty was crumbling from within.

The Han had certainly had its critics, as well as its champions. A long-standing complaint against the imperial administration was that nepotism was rife, and that those who were truly talented were not being chosen for positions of importance in the government, instead being passed over for insiders, friends, and the ingratiating, often who were incompetent hacks. If this was so, we can imagine that the administrative work of empire would have suffered, groaning under the weight of the incredible task of governing an empire and depending on such inadequate people to support it. It could, of course, be that those inclined to (or with the time to) write treatises were the spurned and not the successful, who would certainly have been biased and bitter. But this does not explain the large number of accounts we find beginning in the Later Han that explicitly address this issue. Presumably there had always been people rejected and angry, but they did not always air these grievances in print. Something was different about the Han that led people to do so. It also cannot be said that the men who wrote these treatises were simply unqualified and talentless people who should not have gained position in the Han court. The works they wrote were works of genius, and far outlasted the later Han dynasty. We cannot dismiss the works of such people, works which discussed the most difficult points of ethics, politics, metaphysics, and the sciences, as merely the ranting or complaints of unsuccessful applicants to the Han court. If these clearly brilliant men were not selected, we should instead wonder "what in the world was wrong with the Han?"

It turns out there was quite a lot wrong with the Han. While the beginning of the end-period of the Han is traditionally dated to 189 CE, with the accession of the final emperor of the Han dynasty, Xian, the Han decline had its roots much earlier, and began earlier as well. It was already underway by the time the great comet appeared in 178. By 220 CE, the empire of the Han would be no more, dissolving with the abdication of Emperor Xian, even though it had been dead on its feet, alive only in name and symbol, for some time before. One of the most famous periods of Chinese history, one that has been immortalized in semi-fictional prose, has its roots here—the Three Kingdoms Period.

In 178, Emperor Ling had been in power for ten years, having acceded to the throne following the death of Emperor Huan. He was not a son of Huan, who had no heir, but rather the great-great-grandson of a previous ruler, Emperor Zhang. He was installed on the throne in 168 CE, when he was 12 years old, making him 22 when the 178 comet appeared. He was to die at the young age of 32, in the spring of 189. In the middle of Ling's reign, the empire was beset by a series of rebellions, the most famous of which was the massive and bloody Yellow Turban rebellion of 184, which also plays the role of the opening event of the one of the most famous novels of all time, Luo Guanzhong's *Sanguo yanyi* ("Epic of the Three Kingdoms"). Although Han forces ultimately stamped out the rebellion, it left them significantly weakened in terms of decentralization of power. Rebellion started in the middle of Ling's reign for a number of reasons, but one of the main reasons was the increasing corruption of the regime and alienation of the people from it. Emperor Ling proved the critics right about the nepotism and ineptitude of his regime in the year of the comet, 178, when he began selling offices to the highest bidders. It would not be the last time in Chinese history such a thing happened—but each time, it signaled a hopeless decline. Another major issue, also at the heart of the causes of the rebellions, was the rising power and influence of the eunuchs at court, whose role became almost central. This culminated in 178 with the eunuchs' successful plot to oust Emperor Ling's wife, who feared that she might undermine their influence. Through their intrigue, they drew up an accusation that she was plotting against Emperor Ling using the dark arts, and the hapless Ling fell for their scheme. His wife was imprisoned, which was inevitably a death sentence, and died in captivity.

The court astronomers were learned people, and would have seen, in the crucial year 178, that the end was near. They must have read the visit of this comet as the signal of doom. But who could tell the Emperor? Who could stop the fall? It was likely too late. And more importantly—who would have *wanted* to stop it? The Han court was in disarray, even its own members and closest confidantes losing the faith. Ultimately, the rebellions would continue into Emperor Ling's late reign, and when he died at the surprisingly young age of 32 in 189, of unknown causes (perhaps he was helped?), the empire quickly dissolved. Amidst the rebellion throughout the empire, a lengthy and vitriolic succession crisis was the last thing the frail Han court needed, especially one in which the hated and entrenched court eunuchs would play a major role. But that's what they got. And with the accession of Ling's son Liu Bian, who became known as Emperor Shao for a brief moment, the crisis escalated, with Shao removed by outside parties in favor of his brother, and the inner intrigues of the Han court becoming increasingly insular and ignorant of the massive conflict and crises sitting on their doorstep about to destroy them. There seems to be something in human nature that leads us to be most ignorant when we are about to perish—a protection mechanism that makes us think that if we act like everything is okay, then it somehow will be. Things were not okay for the Han dynasty. The rebellions in the countryside grew into mass movements, factions began to take control of large swaths of the empire. By the early 190s, the Han existed in name only, its emperor being nothing more than a puppet and figurehead, and by 220 CE the Han was no more.

Just as one had with Wang Mang in 23 CE, a comet told the story of the demise of the Han. But, as was only proper, the comet marking the end of the Han was a truly great comet, a blazing trail in the sky that would have been noted by people across the globe. The comet of 23 was likely a small and dim object that only the closest skywatchers, the court astronomers, would have seen. A small comet for a short and unlucky regime. A dynasty as magnificent as the later Han had been, and its predecessor the former Han before it, certainly deserved a once-in-a-lifetime comet to mark its close.

Of all the accounts of astronomical events and celestial objects in the world, that of the Chinese is by far the longest and the richest. There are many events for which the Chinese record is the only existing. There are many events that we can now predict must have happened in the past, especially eclipses and comets, and for many of these hypothesized past events, there have been Chinese records (this can only have been the case for those eclipses and comets visible from China, of course). The supernova of 1054, for example, as discussed at length in the last chapter, was first seen and documented by Chinese astronomers. There are other documented supernovae for which the only existing documentation is that of the Chinese, as well as many comets of this type. Comets, as we have seen, could hold great human significance for the Chinese, sometimes signaling the doom of a dynasty, and sometimes its glorious rise or continuation. But what was the nature of these odd "broom stars" the Chinese so diligently noted and followed?

References

Didier, John, 2009, "In and Outside the Square: The Sky and the Power of Belief in Ancient China and the World, c. 4500 BC- AD 200", 3 Volumes. *Sino-Platonic Papers* 192.

Krupp, E.C., 2003, *Echoes of the Ancient Skies: The Astronomy of Lost Civilizations* (Mineola, NY: Dover Publications)

Lee, Janghee, 2005, *Xunzi and Early Chinese Naturalism* (Albany: SUNY Press)

Loewe, Michael, 2006, *The Government of the Qin and Han Empires: 221 BCE- 220 CE*, (Indianapolis: Hackett)

Needham, Joseph and Wang Ling, 1959, *Science and Civilisation in China: Volume 3* (Cambridge: Cambridge University Press)

Sanft, Charles, 2014, *Communication and Cooperation in Early China: Publicizing the Qin Dynasty* (Albany: SUNY Press)

Steinhardt, Nancy, 1990, *Chinese Imperial City Planning* (Honolulu: University of Hawaii Press)

Thomsen, Rudi, 1988, *Ambition and Confucianism: A Biography of Wang Mang* (Aarhus: Aarhus University Press)

Xu, Zhentao, David Pankenier and Taotiao Jiang, 2000, *East Asian Archaeoastronomy: Historical Records of Astronomical Observations of China, Japan, and Korea* (Amsterdam: Gordon and Breach Science Publishers)

Chapter 5
Comet—A Sea of Dust and Ice

What was it that the Han court astronomers saw in the sky in 178? What did they understand about the phenomenon of the "broom star", and what do we know about these events today? The Chinese certainly understood that these strange objects were relatively close by in our celestial neighborhood, as they moved quickly across the sky, similar to the motions of the planets, rather than the distant stars, which do not appear to us to move through orbits (although we know today that they do). Although their rate of motion was similar to that of the stars, one thing Chinese astronomers noticed was that the *direction* and plane of motion of comets was not often the same as that of the planets. Comets could come from any direction, and generally did not move along the plane of the planetary orbits, which we call the ecliptic. A comet could appear high above the ecliptic and move down across it, or it could appear near it and move in the opposite direction of the planetary motions. The paths of comets were not predictable in the way that those of the planets were, or the sun and the moon were. This unpredictability made them the perfect divinatory symbol. The appearance and motion of comets were not predetermined, not regular and necessary. They instead signaled (and illustrated) a change of the natural order, a critical shift in nature, for good or evil.

Of course, we know today that the comets actually *do* follow a patterned order (for the most part), that their orbits and motions are just as much determined by the regular patterns of the laws of physics as are the planets, sun, and moon. Comets do not, indeed, act like the other more familiar objects of the solar system, in part because comets are the rogues of the evolutionary process of the creation of the solar system.

At the beginnings of the formation of the solar system, the material that condensed into the sun left behind, if you will, a bit of material that is sizable by our standards but ultimately tiny by solar standards. The matter that condensed into the central star became the sun—all of this matter was in rotation, and thus the sun was born already rotating. Indeed, as the matter condensed via gravity and became the

A. McLeod, *Astronomy in the Ancient World*,
Historical & Cultural Astronomy, DOI 10.1007/978-3-319-23600-1_5

star that is our sun, its rate of rotation increased, due to a law of physics we know as the conservation of angular momentum (a law which also explains why figure skaters spin faster when they draw their arms into their sides, for example). The leftover material from this process, also rotating now around the sun, cooled in the cold of space and became rock and ice. Over time, the rock and ice in their orbits around the sun collected together, due to their motions and to gravity, into larger objects. Rocky objects collided with one another to form new objects, and these larger objects attracted yet more matter due to their increased gravitational force. Eventually, this process led to the development of the planets. As we know, the interior planets, from Mercury to Mars, are primarily rocky objects, most of which (except Mercury which is too small) have outer layers of gaseous atmospheres, created by the venting of gases created within the planets, as well as the evaporation of water. These planets, of course, are large enough that they are not *just* large rocks. At a certain level of pressure, solids like rock and metal will inevitably melt, and the planets have accrued enough matter that the vast majority of the interior of the planets is not solid, but molten rock. It is in part due to the outgassing of this molten rock and the processes happening within the Earth that our own planet has an atmosphere (and this process is also behind plate tectonics, which is responsible for volcanoes, earthquakes, mountains, oceans, and most of the features on the Earth's surface), and luckily this planet is also large enough (unlike Mercury and Mars) that the atmosphere does not just fizzle off into space. While the interior four planets are rock-based, the outer planets, which we collectively refer to as the "Gas Giants", have a much different constitution. These bodies are made primarily of gas and ice, with much less rocky material than their enormous size would suggest, and thus operate in many ways differently from the rocky planets. In some ways, the Gas Giants are more akin to the sun than they are to the interior planets, as the sun is also a dense gaseous object. Unlike the sun, there is no nuclear fusion taking place at the core of the Gas Giants, as they do not have enough mass for this to take place. If something were to increase the mass of one of the planets, at some point it would reach a critical mass at which the pressure at the core of the planet would reach such a point that the planet would indeed become a star. This is, indeed, exactly how stars form, and our sun formed in the same way. There is some indication that Jupiter has a mass that is not all that far from the critical mass needed to start the fusion process of creating a star. Indeed, Jupiter's (relative to us, at least) enormous mass is high enough that there is a process of radiation at the center of the planet that makes Jupiter the only planet of our solar system that radiates more energy than it receives from the sun. Jupiter is a near-star, just missing the threshold of mass that would have made it one. Our solar system could well have developed in such a way as to include two stars—our sun and Jupiter would have formed a binary star system similar to our neighboring binary Alpha Centauri (comprised of two stars), and we would not be here to talk about it.

Although most of the material in the solar system either became part of the sun or one of the planets (and the majority of this is in the Gas Giants, especially Jupiter, the "almost-star"), some of the rogue material that did not make it into the sun or planets scattered throughout the solar system in orbital spheres around the sun. To give a sense of the amount of such material, consider the mass of the solar system. 99.8 % of all the solar system's mass is contained in the sun, leaving the planets as the real "remnants" of the solar construction process. Of the remaining 0.2 %, Jupiter makes up 71 % and Saturn 21 %, thus collectively comprising 0.184 of the solar system's mass, leaving all of the other planets, including the two remaining Gas Giants (which take up the majority of the remaining mass) and all of the rocky planets, plus the comets and asteroids, to comprise only 0.016 of the solar system's mass. Our own Earth comprises a miniscule percentage of the total mass of the solar system, and the entirety of all cometary and asteroid material in our solar system makes up a miniscule amount of material even in comparison to our tiny earth. All of the extra-planetary material in our solar system makes up around the same mass as a mere 4 % of our moon. Comets and asteroids pack a punch far above what their weight class would suggest.

It may be hard to believe that such truly miniscule objects, sometimes no larger than a football stadium or even a large truck, could play such a large role in human astronomy, in the development of the solar system, and certainly in the evolution and fate of life on Earth. One crucial question to answer concerns comets: why, if they are such tiny objects, do they appear as the most massive and brilliant objects in our skies when they appear—sometimes so bright they are visible even during the day? There are perfectly good and very interesting reasons for this. But first, we have to consider a distinction between three different kinds of objects comprised of this extra-planetary solar system rock, only one of which shows up in our night sky in the brilliant display put on by "broom stars" like the great comet of 178 CE.

All kinds of non-planetary objects inhabit our solar system, from massive asteroids and comets to tiny specks of dust, rock, or ice of the kind that forms the enormous belt system around Saturn. Two objects in particular are relevant for our purposes here: asteroids and comets. The difference between the two is not very great, but even this slight difference is enough to make the difference between an all-but-invisible object and one that lights up huge swaths of the sky. Asteroids are primarily rocky objects, generally orbiting the sun in one of two orbital rings around the sun, the earliest known of which is simply called the "asteroid belt", a ring of orbiting asteroids around the sun between the orbits of Mars and Jupiter. The other primary source of asteroids is in a more distant belt called the Kuiper Belt, after astronomer Gerard Kuiper, who hypothesized its existence, but did live to see the discovery of the belt, as he died in 1973. Ultimately, verification of its existence would have to wait until the development of more sophisticated telescopic equipment, and came only in 1992. The Kuiper belt is far outside the orbit of the planets, being as distant from the orbit of Neptune as about two-thirds of the distance between Neptune and the Sun. Neptune is 30 AU distant from the sun, while the Kuiper Belt is 50 AU. For comparison, the Earth is 1 AU from the sun (the earth-sun distance is the standard measure for the AU). These rocky objects are part of what comprises the Kuiper Belt, but they exist there along with a different kind of object, which is our main concern—comets.

Comets are similar to asteroids in a number of ways. Generally, the size of comets and asteroids, as well as their irregular shapes, are similar. The classic image of the comet or asteroid, looking like a massive rock flying through space, can be explained by gravity. In larger objects like planets, gravitational force molds the objects into spheres, as the (relatively) large mass toward the center of the object exerts enough force on the area toward the surface to deform the rock and/or ice. In the case of smaller objects such as comets or asteroids, the gravitational force is not great enough to overcome the integrity of the rock or ice, and thus these objects retain their "rock-like" irregular shapes. There are some objects in the Kuiper Belt that are large enough that they have taken on spherical shape, and many of these objects have become classified as *dwarf planets*. The official definition of a dwarf planet by the International Astronomical Union makes the shaping of the object by gravity part of the definition, with the addition that such objects are *not* large enough to clear their orbital paths of other small objects, such as comets and asteroids.[1] Thus, such tiny dwarf planets as Ceres, Makemake, and Pluto (which used to of course be referred to as a planet) inhabit the Kuiper Belt along with comets and asteroids. Comets are small enough to have not formed into dwarf planets, and unlike asteroids, they are primarily made of ice and dust, with smaller rocky cores. One may think of them like asteroids with a thick outer crust of ice and dust. Such objects exist in the Kuiper Belt as well as even more distantly in what is referred to as the Oort Cloud, which is a spherical region of objects in orbit around the sun, at the very outer reaches of our solar system.

Most comets spend their existence within regions like the Kuiper Belt and the Oort Cloud, and are destined to continue in their orbits around the Sun for the rest of their existence. But those comets that we know, the comets that enter into human history through documents like those of the Chinese in 178 CE, or the famous "Caesar's Comet" of 44 BCE, are ones that have been shaken from their cometary slumber, through the gravitational effect of some passing object such as one of the minor planets or a stronger tugging from a particular passing of multiple planets. These comets fall toward the sun, their previous orbits altered so as to cause a closer approach to the sun than they would otherwise have made. The comets that we know, the ones that appear in our skies, are a subset of these comets—ones that approach close enough to the sun that they come within the orbits of the planets. Close enough that the effects we witness in the sky as the "broom star" phenomenon can be seen. There are a couple of different kinds of orbits of such comets.

There are those comets, such as the famous Halley's Comet, which we refer to as "Short-Period" comets—comets that come in for a close pass into the inner solar system before launching on their long elliptical orbits back out into the distant solar system and then returning. Halley's Comet itself has an orbital period of 75 years, appearing in the night sky about once during one's lifetime—and if one is lucky, twice. The last appearance of Halley's Comet in the inner solar system was in early

[1]The IAU's Resolution B5 ("Definition of a Planet in the Solar System") reads: "A planet is a celestial body that (a) is in orbit around the sun, (b) has sufficient mass for its self-gravity to overcome rigid body forces so that it assumes a hydrostatic equilibrium (nearly round) shape, and (c) has cleared the neighborhood around its orbit."

1986—an event I witnessed as an almost eight-year-old child with my father. It's an event I hope to be around as an 83-year-old in 2061 to see—especially because the 2061 appearance is supposed to be much better than was the 1986 appearance, one of the least impressive appearances of the comet, in part due to its distance (0.42 AU at its closest). Short period comets generally originate in the (relatively) closer Kuiper Belt, and have orbits that take them into the inner solar system and back out toward their origin points before returning (Fig. 5.1).

The other main type of comet is the "Long-Period" (or non-periodic) comet, which, as its name suggests, originates further out in the solar system, mostly in the Oort Cloud, and also perhaps occasionally even further afield. The category of Long-Period comets is a more internally diverse one than is that of the Short-Period comets. The category is basically a "catch-all" for those comets that do not qualify as Short-Period comets–and in this way, the term "Non-Periodic" perhaps better

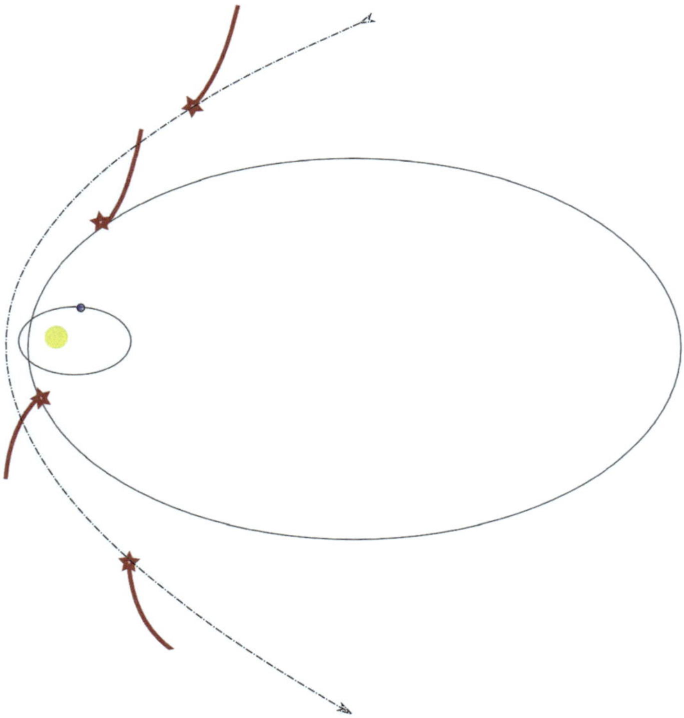

Fig. 5.1 Comets of different periods. The *inner path* describes the orbit of a *short period* comet, with orbits around the sun that bring them into the inner solar system more regularly, in cycles of less than 200 years. One of the most famous comets in history, Halley's Comet, is a short period comet, appearing every 75 years. Its next scheduled appearance is in 2061. The *outer path* describes the orbit of a *long period* comet. Comets in this category vary. They range from comets orbiting the sun with periods longer than 200 years (a 300 year comet falls into this category), to comets that do not orbit the sun and will only have a single appearance in the solar system (a comet that is pulled from an asteroid belt and escapes the solar system after is pass of the sun) (Illustration by the author)

captures the nature of this category. Some of these objects have been observed only once, and thus must have periods over a few hundred years, or none at all. Other objects are observed to have an orbit that is nearly parabolic during its pass through the inner solar-system, and so its orbit is either so long that it cannot be determined, or else the object may be on a single pass through the inner solar system, being flung out into interstellar space (or wherever it may be captured by the gravity of another object) rather than kept within orbit around the sun. We can sometimes distinguish between these two types of comets—those with enormously long periods and those without orbital periods that will continue on a trajectory outside the solar system unless they are trapped before their exit.

Within both of these categories, there is another interesting type of comet–the *sungrazing* comet. Such comets, on their pass through the inner solar system, pass so close to the sun that they are sometimes burned up before their passage around it. One class of such sungrazing comets, the *Kreutz sungrazers*, are parts of an original larger comet that broke up into multiple parts. Such comets, for reasons to be discussed below, are often the brightest and most spectacular of comets to appear in our skies, if they survive their near passage to the sun.

Comets, as comprised primarily of dust and ice, are frozen solid in their distant home in the Kuiper Belt, the Oort Cloud, or even further afield. The dust and ice is compacted together with the rock, in a combination sometimes referred to as a being akin to a "dirty snowball" (or perhaps a better image is a "dirty iceball"—of the kind that sometimes gets thrown at a person one *really* dislikes in a snowball fight). Because of their distance from the sun, like other icy objects such as Pluto (which might be thought of as a comet so large that it became a dwarf planet) keep all of their ice and dust frozen to their surfaces. When their orbits are disrupted and they are sent careening into the inner solar system, this situation begins to change. As they approach the sun, the increased light and heat ionizes the ice and dust, causing a gaseous "atmosphere" to form around the comet, which is referred to as the *coma* of a comet. It is an atmosphere in only a temporary sense, as the outgassing of a comet could not create a layer of gas that could remain on a comet, which is too small to capture an atmosphere. If the comet were to continue its trip for the rest of its life through the inner solar system, the coma would quickly dissipate from the small object. Because of this lack of strong gravity, the coma of a comet can become far larger than the comet itself. When we see a comet in the sky, the central conical or semi-spherical looking object at the head of the tail is actually the coma, not the comet itself. The comet itself is far too small for us to see directly. In a sense, we never actually see the nucleus of a comet, that is, the comet itself, we only see its effects.

The second major feature of comets, the *tail* effect, is explained by the interaction of the coma with the solar wind. The solar wind is a steady stream of particles released by the sun, both protons and electrons (a more violent and dramatic outburst of which happens during a solar flare). The solar wind, as emitted from the entire surface of the sun, travels outward from the sun in all directions into the solar system. Contact with the solar wind, which is heavier in the inner solar system than it is at more distant locations like the Kuiper Belt, pushes the ionized gas and dust particles in the coma away in the direction of the solar wind. Thus, we see the long, bright tails streaming from comets in the sky. In addition, you may notice two distinct tails on a comet (depending on its makeup)—mostly a long white or yellow tail, and a blue one, trailing in a different direction. What explains this is the difference in the solar wind's effects on the two different ingredients making up the coma of the comet. Dust particles are responsible for the white or yellow tail, and because dust particles are heavier than those of the ionized gas that forms the second tail, they are pushed more slowly, and generally this tail is arced slightly away from the straighter blue tail. The light ionized gas is carried quickly by the solar wind, and often appears as an arrow straight line shooting from the comet (Fig. 5.2).

One may notice, as many ancient astronomers did, including the Chinese, that the tails of comets are always facing away from the sun. It is the solar wind effect that creates the tails of comets discussed above that is responsible for this. Thus, it is not correct to say that the tail of a comet is coming out from *behind* a comet. A comet does not experience the effect of wind in an atmosphere in the way we would experience it on earth. Usually, when we see things streaming behind a moving object, it is explained by their direction of movement. The reason for this is that our movement through the air, which is all around us, creates an artificial wind, moving in the direction opposite to our motion. This is why we need windshields on cars, for example. In space, however, there is no atmosphere, and thus no artificial

Fig. 5.2 Comet Hale-Bopp (C/1995 O1) was one of the brightest comets of modern times seen from earth, and reached perihelion in 1997. This photo of the comet offers a clear view of the two distinct tails of a comet. The *blue tail* on top is the *ion tail*, while the *dust tail* trails below. Both tails are explained by interaction of the comet with the sun. The ion tail is created by ionization of gas in the coma of the comet, which forms as gas is released from the heating of the comet by the sun. The solar wind pushes this ionized gas from the comet in the direction away from the sun. The dust tail is created from dust particles in the coma, which are pushed by pressure from sunlight away from the comet. The ion tail always points directly away from the sun, while the dust tail is determined in part by the direction of orbit of the comet (Photo credit: NASA)

wind created by motion. Thus, the ionized gas and dust released from a comet are not affected by the motion of the comet, but only by the solar wind. The motion of the comet can be, and often is, in a completely different direction than away from the line of the tail. It could be moving perpendicular to the tail, or, indeed, if it is moving directly away from the sun on its way back out into the dark reaches of the solar system or beyond, it could be moving *with* the tail. So in no sense should the tail be thought of as behind the comet. There is no way of knowing, from seeing a comet at an instant, what its direction of motion is. All we can know from a snapshot is roughly what direction the sun is in.

Comets, as a celestial phenomenon, are much more commonly seen than supernovae, even though as a whole there are many more detectable supernovae today than there are visible comets—that is, ionized comets that enter the inner solar system. Part of the reason for this, of course, is that the only supernovae humans have been able to see until relatively recently in our history have been those happening in our own galaxy, which happens only at the rate of about two per decade, as mentioned in the previous chapter (although we haven't been lucky enough to have one for over 400 years, since 1605). Visible comets appear much more regularly. There have been numerous visible comets in the last 30 years, for example, including magnificent ones like Halley's Comet in 1985, Hale-Bopp in 1995, Hyakutake in 1996, and smaller appearances like the more recent PanSTARRS comet in 2012. In the ancient world, those who watched the skies closely, such as court astronomers in China, documented numerous comets, including many likely missed by most of the population. Indeed, some of the most important celestial events in the history of humanity have been comets.

Historical Comets and Their Effects on Culture

Of the many famous comets to appear in the skies documented by humans throughout history, there are a number of notable ones that changed civilization. A number of the appearances of Halley's Comet, which enters the inner solar system every 75 years, have been thus important, although no one knew that these comets were actually one and the same object until the 17th century, with the work of the English astronomer Edmund Halley. There had been a number of notable passes of Halley's Comet in human history. Among these were that of 837 CE, which represented one of the closest passes of a comet to the Earth in human history, as Halley's passed a mere 0.03 AU from the Earth at its closest during this passing. This is closer to the Earth than its nearest neighbors, Venus and Mars, ever come. Another notable pass of Halley's Comet came in 1066, the year of the Norman conquest of the British Isles. According to legend, it appeared before the battle of Hastings, and signaled William the Conqueror to launch his ultimately successful invasion, which certainly changed the course of history from what it otherwise would have been. The appearance of the comet in 1759 marked a triumph for Edmund Halley's theory of the return of the comet, but unfortunately one he did not live to see, happening as it did 17 years after his death. Short period comets such as Halley's Comet lose a bit of material every time they pass into the inner solar system, through the outgassing process that causes the coma and the tail, and one day over many passes, the comet is destined to eventually dissipate and come apart, losing all of its material. It loses only a tiny percentage of its mass to this process with each pass, so it will take many thousands of years for this to happen. Although we seem to see in the sky an amazing profusion of particles blown from the comet, because the reflectivity of the ionized gas and dust is so high, the amount of actual material we are seeing is relatively small, and tiny in comparison the mass of the comet.

Other famous comets include, of course, the 1577 comet that Tycho Brahe used to demonstrate in Europe the inadequacy of the Aristotelian/Ptolemaic view of the perfection of the spheres—something that the Chinese were already well aware of, and had been for thousands of years. In the ancient world, there were many notable comets, such as "Aristotle's Comet" of 372 BCE, which was documented by the philosopher in his *Meteorologica* as a particularly great comet—even though Aristotle did not properly understand the nature of comets, thinking that they were phenomena within the atmosphere of the Earth, and not within the realm of the spheres. "Caesar's Comet" of 44 BCE appeared in the skies over Rome (and elsewhere in the world, of course) after Julius Caesar was assassinated by conspirators of the Senate, and was taken by the common people to represent the spirit of the mighty Caesar ascending to the abode of the immortals.[2] The comet became part of the imagery in later years of the divinity of Caesar. This contributed, of course, to the deification of Caesar that his assassins had been so concerned to avoid, and eventually to the inevitable fall of the Republic and the rise of the figure of the Emperor. Yet Caesar's heir, Gaius Octavius, who would become the first Roman emperor Augustus, had plenty of reason to encourage this association. Perhaps ironically, Augustus' own death, in 14 CE, also coincided with the appearance of a red-colored comet.[3] The comet associated with the Aztec ruler Moctezuma II in 1517 just before the Spanish conquest played a large role in the later legends associated with him. In this case, as in that of early China, the appearance of a comet was seen as a sign of doom. It was of course not always so, as we see in the Roman case.

Even if appearances of comets never themselves played major roles in directing the course of history, they are often *interpreted* as playing such a role by those looking to the past for answers. The true causes of events like the fall of the Aztec empire, the collapse of the Han dynasty, or the rise of the Roman empire are many and complex. Historians, archaeologists, and others debate these issues to this day. Comets serve as a very visible and immediate option to answer these questions.

---

[2]Schechner (1997, 24-25).

[3]Rehak (2006, 71), basing his claims on Cassius Dio, says that *several* red comets appeared in the year of Augustus' death, which according to Cassius "confirmed his deification".

They appeal to our sense of mystery, drama, and desire for simplicity. It is easy to see why many people in numerous societies would have seen them as signs or even causes of momentous events. Comets are sufficiently rare, startling, and enormous—more striking in many ways than supernovae. Although most comets visible from the Earth are not nearly as bright as a visible supernova, the brightest comets far outshine visible supernovae in apparent magnitude. One must be careful to refer to *apparent* magnitude here, because the light emitted from a supernova is many billions of times greater than that reflected from a comet. It is the enormous distance between us and supernovae that account for their relative dimness as measured against the brightest comets. Were these supernovae as close as the comets we see in the sky, we would not be around to see their brightness, because the Earth would be near the center of the massive explosion. If our nearest stellar neighbor, Proxima Centauri (4.24 light years away) were to become even a relatively dim supernova (which is impossible given that it is a red dwarf), it would shine far brighter than the sun, and would also burn away our atmosphere.

The Philosophy of Unification in Early China

The Chinese account of the connection between comets and other celestial phenomena and the human world did not grow out of a naivete about the world, nor did it appear overnight. The ancient Chinese recognized certain important continuities between the natural world and the human world, and thought that these regularities could be explained by observing the patterns and discontinuities of each of these worlds. It would be arrogant of us to reject or ridicule the early Chinese notion that the human realm might be affected by celestial events such as comets and supernovae, or meteorological events. Especially if we do not know what the position amounts to.

The theory behind the ancient Chinese view of the resonance or unity between nature and humanity is a complex one, and one that developed and was made more explicit and clear over time. We see in the earliest known Chinese writings, inscriptions on oracle turtle-shell bones meant to tell the future, that the Chinese engaged in divinatory practices from the dawn of civilization, well before any worked out theory of the connection between nature and humanity had been formed.[4] Such views were implicit in the practices and worldviews of these ancient peoples, and what we see developed in the late Warring States and early Han is simply a way to develop what is already inherent in these views.

One major issue that contributed to the expression of an explicit philosophy of resonance was that of truth and methodology, and a new understanding of the best methods for uncovering truth, developed in the late Warring States and early Han.

---

[4]Lisa Raphals considers the questions of the purpose of divination (Raphals 2013, 2-9) as well as specific Chinese practices including oracle bone reading (Raphals 2013, 128-147).

In texts such as the *Lushi Chunqiu* ("Spring and Autumn Annals of the Lu Family") and *Huainanzi* ("[Book of] The Master of Huai Nan"), the idea of the combination or syncretism of various areas of thought (including positions that appear on the surface to be mutually exclusive) is presented. Both texts attempt to create theoretical structures under which a synthesis of all human knowledge can be generated. In the *Huainanzi* in particular, the concept of *dao* ("the Way") understood in a naturalistic sense, is that under which all things are organized and synthesized. The *Huainanzi* uses the distinction between root (*ben*) and branches (*mo*) to explain the organization of all aspects of the world within the single *dao*. The *dao* is the root, while the particular schools, teachings, and areas of knowledge are the branches. The primary purpose of the *Huainanzi* may have been to give instruction in rulership (the good ruler is one who knows the root—the *dao*—and thus understands how to employ or use the branches). Still, this idea of the correspondence of all things and all areas of knowledge to an overarching Way in which they are synthesized made sense of understandings of the non-human world, including the celestial realm, as closely related to the human world. The *dao* that unifies them all is present in some sense in all of these disparate realms. Since this is the case, we might say that the same pattern that reveals itself in the unfolding of events in the celestial (or any other) realm also reveals itself in the human realm. Later philosophers in the 11th century would discuss this idea using the concept of *li* ("principle", "pattern").[5]

While these views may seem foreign to Western readers, they actually closely parallel some ideas behind modern naturalism and science. We hold that the laws of physics work the same way for everything that exists, and thus ultimately the motions of the planets and the formation of galaxies operate through the same principles as those manifested through human activity. We might say that these principles, these laws of nature, are *manifested* by the actions of stars and the actions of humans both, albeit in different ways. And, just as is the case with the *dao* in *Huainanzi*, if we understand the physical laws and all their implications well enough, we should understand just how events in every part of the world–solar, geological, or human—will play out. The scientific naturalist (at least one who is reductionist) is committed to this.

The *Tianwen* chapter of the *Huainanzi* is devoted specifically to astronomy. It explains the methods of early astronomers, as well as how astronomy fits into the overall plan of the *Huainanzi* of revealing the *dao*.[6] One passage in the *Tianwen* chapter describes a number of continuities between Heaven (*tian*), here thought of as the celestial realm, and humanity:

---

[5]Specifically the Song "Neo-Confucian" philosopher Zhu Xi is associated with this concept, which was at the core of both his metaphysics and ethical theory.

[6]John Major writes: "The principal message of the chapter is that all things in the cosmos are interconnected, that human plans and intentions are subject to the influence of various cosmic cycles and correlations, and that such cycles and correlations can be understood and taken into account in the formulation of policy." (Major et al. 2010, 109).

> Of all the creatures that move and breathe, none is more prized than humans. The bodily orifices, limbs, and trunk all communicate with Heaven:
>
> > Heaven has nine layers; man also has nine orifices.
> > Heaven has four seasons, to regulate the twelve months;
> > Man also has four limbs, to control the twelve joints.
> > Heaven has twelve months, to regulate the 360 days;
> > Man also has twelve joints, to regulate the 360 nodes. (Major, Queen, Meyer, and Roth, 2010, 143–144)

The resonance between humanity and nature entailed that humanity should be considered part of or necessarily linked to the processes of nature. Some have argued that the Chinese, beginning with the early philosophical material, have been markedly "this-worldly" in their thinking, neglecting consideration of transcendent entities inhabiting other worlds or realms, something we see explicitly discussed in Western thought at least as far back as Plato.[7] While this is perhaps true, it can also be misleading. The concern with this world we see in early China is not a rejection of non-human entities such as spirits and other agencies involved in influencing the world, in some cases even a "divine" agent (as in the case of the Mohist *tian*, "God, heaven"). The early Chinese did not see these agents (*shen*, "spirits", *tian*, "heaven") as transcendent or outside of *this* world, but rather as within it and a part of it, just as are humans.[8] One may argue that the Confucian resistance to discuss issues of heaven and the spirits and instead focus on human issues was a rejection of supernaturalism, but this would be a mistake. The Confucians, unlike other schools such as Daoists and the related Zhuangists, saw the specifically human, the *ren dao* (way of humanity) as uniquely and exclusively worthy of our attention. In one famous passage in the *Analects*, Conficius also tells a proto-Daoist hermit "we cannot follow the way of the birds and beasts." This is of course not intended as a claim that the way of the birds and beasts somehow belongs to a different realm or world than that of humanity. Confucius is simply explaining that we should be concerned with the way of humanity and not with that of birds and beasts. And the same thing follows for the way of the ghosts and spirits, and the way of *tian* (Heaven). These are different kinds of thing, and each has a different *dao* (way). But all of these beings and all of these *dao* are within this world. This is why one does not see any claims in Chinese cosmogonies that *tian* (Heaven) or some other transcendent entity pre-existed the cosmos, as we see in Western (Judeo-Christian-Islamic) theology.

The early Chinese recognized, as have all people in human history, the numerous connections between the non-human realm and human life, our fate, behavior, and development. In addition, different non-human realms affect one another, which in turn

---

[7]Roger Ames is a well-known proponent of this position, in particular that there is a lack of the idea of *transcendence* in early Chinese thought. Ames argues for this in many (indeed almost all) of his works, but a good and concise statement can be found in the introduction to (Ames and Rosemont 1998).

[8]The question of the relationship between humans and these other beings is not so easy to answer. Michael Puett argues against the idea adopted by a number of major scholars including Ames, earlier scholars such as A.C. Graham, Benjamin Schwartz, and others (Puett 2002).

affect us. The change of seasons affects the growth or decay of plants, which influences our lives. Earthquakes and drought cause changes in the human realm and in that of the "birds and beasts" alike, and the actions of the birds and beasts can have significant influence on the human realm. None of the beings and realms in the world, even though they may be different, are completely independent of one another. Since the early Chinese recognized the spirits, Heaven, and other such entities as distinct entities in the same world as humans, the birds and beasts, and all other things, it followed that these beings could have an affect on the human realm as well. It is even more apparent that the celestial realm, including the sun, moon, planets, and stars, can influence the human world. The sun creates warmth, influences the growth of plants, the progression of the seasons. The moon's light creates changes in the animal and human world, controls tides, and has a number of other effects the early Chinese did not recognize, including causing precession of the equinoxes. It was a short step for the early Chinese to make to consider the planets and stars (especially the circumpolar stars) and phenomena happening among these stars as having some effect in the human realm, as well as signaling some corresponding phenomenon in the human realm.

But how do we get to a view of celestial events as portending significant events in the human realm such as the rise and decline of dynasties or rules, drought or plenty, or the proper time for a transition of rule period? All cultures have recognized the continuities between and reciprocal effects of interaction between nature and humanity. But the Chinese transformed and codified this into an elaborate system of interpreting interactions between the sky and humanity in specific and detailed ways. They formed an observational and interpretive astronomical system more central to their intellectual culture than perhaps in any other society except that of the Maya, who were equally, or perhaps even more, concerned with astronomy.

The answer has to do with the system of thought developed in early China, in the late Warring States and early Han periods (but implicit far before this), that systematized Chinese understandings of naturalism. This system combined the idea of the syncretism or unification of viewpoints, entities, and events under a single *dao* (Way) or *yi* (one) (as mentioned in the *Huainanzi*), itself a kind of naturalism, with the idea of the mutual influence or resonance of things with one another. The result of this combination was what we today refer to as *correlative cosmology*, which is normally associated most closely with the early Han period (206 BCE–24 CE). Correlative cosmology codified to some extent the connections between different events and entities. The system explains the interaction of things and events and their effect on one another through attributing to all things association with one or more of the *wu xing* ("five phases/processes"). It is important no note here that the word used, *xing* (process) suggests something in motion and changing, an *event* rather than a substance. *Xing* can be used to refer to process, event, and practice. It is not akin to the word that more directly implies "thing" in terms Western thinkers may recognize as substance, *wu*. Of course, *wu* does not specifically refer to anything like a substance either, and the idea of substance in the sense of the Latin *substantia* or Greek *ousia* would be problematic to read into early Chinese thought at all. But certainly *wu* is much closer to this notion than *xing*, which clearly refers specifically to event-like phenomena.

The "five processes" then which all things are composed of or related to are not like basic atoms, in the sense the Presocratic philosophers of ancient Greece thought of them. Sometimes *wu xing* is translated as "Five Elements", drawing a similarity between the Chinese theory and the Greek theory of the "four elements" (*stoicheion*). While there is some resemblance, this identification is misleading. The Chinese theory grounding correlative cosmology is processual and event-based in a way the Greek theory is not, and conflating the two can lead to a deep misunderstanding of the Chinese system of correlative cosmology. If all events and things are comprised somehow of the basic processes, then certain features of the basic processes are relevant in explaining how the things act as they do (so far, we are on shared ground between the Chinese and Greek theories). The processes, similar to the elements of Greek thought in their reference to what were seen as basic *events* underlying all things: fire, water, earth, tree/wood, and metal. The characteristics of each thing are determined not by its being *comprised* of collections of the processes (it is here that we see one difference between the Chinese and Greek systems), but based on *transitions* between different processes. Mutually interactive processes have to give way to one another, just like five musicians on a stage must give and take, and engage in dynamic interaction. They can't simply all play at once, over top of one another. The result will be unintelligible noise. Rather, when one advances, the others retreat toward silence, or pull back a bit to allow the one a space to perform, and then this musician in turn gives way to another. None of the musicians may ever completely *stop* playing, but there is an ebb and flow in their intensity and presence, based on their interactions with the other musicians. Anyone who has seen a jazz show will be familiar with this. This is the same way the processes interact with one another. One "overcomes" another and there is one larger overarching process in which the various sub-processes of the five *xing* take place. We can call this process the *dao* (Way), an idea borrowed from early Daoism via the Han syncretist text *Huainanzi*. *Dao* has many senses in early Chinese thought, and almost every thinker who used the term meant something different by it. In the context of correlative cosmology, we can safely say that *dao* should be considered as the universal or summative process within which all the lower-level processes take place. At the same time, these lower-level processes cannot be considered as completely independent or autonomous from one another or the whole. If we think of large-scale processes like hurricanes, we can distinguish different sub-processes or parts. A hurricane will have an eye as well as arms, but neither of these is independent of the other parts and of the whole hurricane. The eye of a hurricane only exists as the focal point of the rest of the storm, for example. Part of its very identity includes the other parts of the hurricane, and the whole of the hurricane. There is no eye-of-hurricane on its own. Such a concept does not even make sense. Insofar as the eye of a hurricane exists, it exists only relationally, only as an embedded process within a hurricane.

All this is to say that on the correlative cosmological system, the relationships between the five processes are like that between jazz musicians in a performance, or between the eye and arms of a hurricane. They are codependent and mutually responsive. When something happens to one, it causes an effect in the others. When one acts, the others respond. Thus it is with everything in nature, according to the system of correlative cosmology. And since humans are part of this single world process, along with the birds and beasts, the spirits and Heaven, and the planets and stars, all of these things can influence and be influenced by humans. The patterns we observe in the stars resonate with inherent patterns in the human realm.

In later Chinese philosophy, influenced by Buddhism, one sees an extension of the early ideas of correlative cosmology. As is generally the case when systems of thought enter a new cultural context (as with the Indian system of Buddhism entering the very different context of China), in order to understand or make intelligible the new system, similarities between it and native or well-understood systems are found. Interestingly enough, there turned out to be large and important similarities between Buddhism and early Chinese thought. Especially in certain forms of Mahayana Buddhism that became influential in China, particularly that based on the *Avatamsaka Sutra* (Huayan), one finds startling parallels with correlative cosmology. According to Huayan, everything that exists mirrors or includes every other thing that exists. There is no independence or autonomy, rather there is an interpenetration of all things with all other things. This is based on the earlier Indian notion of *satkaryavada*, the view that a cause includes within it all of its effects. The Huayan view expands this, to come up with something much more similar to the correlative cosmological system devised in the Han dynasty. Thus, as Buddhism gained ascendancy in China during certain periods, the correlative cosmological system maintained its influence, and the conception of the sky it grounded remained a central part of Chinese intellectual culture.

Chinese astronomy was one of the most constant features of Chinese intellectual culture throughout most of its history for this reason. It was only with the modern period that Chinese astronomy drastically changed, first with the decline of imperial culture and the slow adoption of Western methods, and then more precipitously with the break from earlier thought in the post-imperial period after 1912 up until the present day. The introduction of Western astronomical thought with the Jesuit missions in the 16th century did not displace distinctively Chinese ways of doing astronomy. The story often goes, concerning Non-Western cultures, that once Western astronomy is introduced, indigenous ways of understanding the cosmos go into decline. This story is clearly false in the case of China and India. In China, much of what was available for astronomical observation when the Jesuits arrived was as good as or better than what the Jesuits brought. It may be an interesting coincidence of history that the early Jesuit missions in China fell at the same time Tycho Brahe was revolutionizing observational astronomy in Europe by utilizing techniques known in China for millennia, particularly the *equatorial* armillary sphere (rather than the zodiac-ecliptic based sphere used in the West since ancient Greek times). As we've seen, the polar and equatorial nature of Chinese astronomy was one of its distinctive features from the beginning, and part of the reason the Chinese were the preeminent observers of the sky for much of human history. The Chinese notions of interconnection of natural processes made the Chinese ways of understanding the connection between humanity and the sky much more natural and intelligible than some of the system later Western thinkers introduced into

China, especially after the time of Newton and the cleft between a deterministic nature and the human (or divine) realm that is made at the dawn of modern science in the West. It is for similar reasons, we will see in the next section, that the introduction of Western astronomical tools and ideas in India did not lead to a decline in earlier Indian ways of thinking about the sky, at least not right away. In both China and India, it was modern crises brought about by colonialism that finally led to the decline and large-scale abandonment of native systems of astronomy and the increasing adoption of western astronomical thought. The relationship of East and West has not always been one of colonized and colonizer. It was not this way in its early history, and things are increasingly moving away from a colonial relationship today as well (though its traces still exist). And it turns out that in the period of exchange between East and West *before* colonialism, the east learned about Western astronomy (and other sciences), but decided that its own systems were preferable.

Why was this? Perhaps it was in part because these different systems of astronomy were based on different ways of thinking about the world, the human place in the world, and different goals based on that understanding. It is not immediately clear or in any way obvious how one *should* understand what it is we ought to be after when we search for the *truth* in the realm of astronomy (or any other science or art). Even if we sweep under the rug the many philosophical difficulties surrounding the concept of truth and take truth to be understood as a kind of correspondence between sentences and the world, such that in astronomy we seek knowledge of which sentences concerning the sky correspond to states of affairs or facts, the question of *which* statements we're concerned with is a non-trivial one, as well as how we determine the truth of these. A statement that the sun is about 71 % hydrogen (by mass) is different than one that the sun represents the continuity of life or that circumpolar motion mirrors the workings of the imperial court. But what makes the first of these a statement of *astronomy* while the second and third are not? They are all three equally about celestial objects, and they are statements that purport to state facts about these objects. The second and third are facts about the connection between a celestial object and humanity, while the first purports to be independent of humanity and about *only* the celestial object itself. But this turns out to be an artificial distinction. All three involve humanity in an ineliminable way. This is one of the features of modern science that has come in for most criticism by theorists, and which shows us the limitations of science at the frontiers. As observers of celestial objects, we cannot view them "from nowhere", and our concepts, our ways of perceiving the world, and our means of knowledge all infuse *what* we take ourselves to know. Modern science has attempted to extract the human from the equation, but this turns out to be impossible. On some interpretations of quantum mechanics, the observer's role is determinative of what is known. Even if one rejects such an interpretation, they generally do it on the grounds of an insistent materialism, which itself must be a background assumption that cannot be demonstrated. We begin to see that there is quite a bit that underlies a scientific conception, or any conception, of the operation of the world. It is neither obvious nor arguable how these basic underlying assumptions should be structured, and most often they are simply inherited from culture or intellectual background. Most of us today tend to think in ways consistent with materialism and Western science, simply because we live in a culture infused with it. Early Chinese astronomers, however, would have no reason to accept these background assumptions rather than those that were deeply entrenched in their own intellectual culture, including correlative cosmology.

If resonance between nature and humanity (*tian ren gan ying*) was at the core of Chinese forms of astronomy, we see something somewhat different in India, a culture that had a great deal of influence on Chinese thought beginning in the Wei-Jin period (3rd century CE), with the rise of Buddhism, and the importation of a number of Indian philosophical concepts and assumptions. One of these central concepts, one that made some inroads in China but never became as established there as it was in India (and for good reason), was that of ritual (particularly Vedic ritual), which serves as a key to understanding astronomical thought in India. While some Indian concepts gained wide acceptance in China through Buddhism, the Indian conception of ritual did not, for a couple main reasons. First, Buddhism itself eschewed much of the traditional Vedic ritual of India, as the Buddhist movement in part grew out of a rejection of or movement away from Vedic ritualism. Second, China itself had a robust ritual tradition, created and maintained by the *ru* (Confucian) scholars. Confucian ritual remained a major feature of Chinese culture throughout its history, and displacing it would have been an insurmountable obstacle for any ideology or school, as Christian missionaries found out for themselves in the 16th–19th centuries.

But enough about impossible tasks—let's now discuss the astronomy of another culture.

References

Ames, Roger and Henry Rosemont, 1998, *The Analects of Confucius: A Philosophical Translation*, (New York: Ballantine)

Major, John, Sarah Queen, Andrew Meyer, and Harold Roth (eds), 2010, *Huainanzi: A Guide to the Theory and Practice of Government in Early Han China*, (New York: Columbia University Press)

Puett, Michael, 2002, *To Become a God: Cosmology, Sacrifice, and Self-Divinization in Early China*, Cambridge MA: Harvard University Press)

Raphals, Lisa, 2013, *Divination and Prediction in Early China and Ancient Greece,* (Cambridge: Cambridge University Press)

Rehak, Paul, 2006, *Imperium and Cosmos: Augustus and the Northern Campus Martius* (Madison: University of Wisconsin Press)

Schechner, Sara, 1997, *Comets, Popular Culture, and the Birth of Modern Cosmology* (Princeton: Princeton University Press)

Seargent, David, 2008, *The Greatest Comets in History: Broom Stars and Celestial Scimitars* (New York: Springer)

Part III
Calculation of Astronomical Wonders

1720 CE

In the desert kingdom of Jaipur, on the brink of war, the Maharaja Jai Singh II likely would have stepped back to consider his legacy. Would he be just another of the numerous and nameless rulers to fall by the wayside of history, ground up under the gears of the Mughal empire, to be cast aside and ultimately forgotten? Posterity would likely not remember him. He had lifted his family out of a few generations of obscurity, debt, and disrepute to get to his current point of magnificence. But his line had been exalted before, yet had still descended into the near oblivion from which Jai Singh II had revived it. Just as all things ultimately decay, he and his line would also one day fade away. There was no way his now rich, but tiny kingdom would be able to resist and maintain forever—it too would fall into empire, of one kind or another.

Amazing celestial events such as comets or supernovae had always been relatively easy to track and to understand, at least on a basic level. Supernovae did not move from the fixed sphere of the stars and comets followed a basic single motion around the sun before disappearing back from whence they came. The motions of the planets and the moon, however, always presented difficulties. In particular, the strange occurrence of eclipses of the sun or moon seemed to happen with a regularity, but one curiously outside the grasp of clear determination. The sun would cover the moon or the moon the sun, in seeming patterns. But the *discovery* of these patterns proved difficult. It was this calculation that concerned Bhaskara II—the calculation of astronomical wonders, such as eclipses or planetary appearances. The elegant mathematics of India could be used to understand these complex motions, to uncover the essential pattern in which the heavens move. In this pattern lays the essence of understanding.

Jai Singh intuitively understood the connection between understanding, modeling the cosmos, and power. A robust ability is to observe and predict motions in the heavens, to track events and motions with the finest possible detail, and to express one's abilities with tools as majestic and awesome in their appearance as they were precise in their use—these were things that displayed true power. While

Jai Singh may have been dispensable as a vassal ruler, he likely thought, he could not have been dispensable as one who understood the essential motions and nature of the heavens, as one who created a ritual model of the universe, a celestial monument that would outlast the prestige and power of his line.

Although many rulers of his time must have thought this way and had these anxieties, Jai Singh II did something truly unique. He did not answer these worries with new palaces, larger harems, or stifling edicts to make himself feel more powerful in the face of ultimate meaningless by making his people suffer. He decided to stake his claim to immortality through the attempt to understand the most mysterious and fundamental of aspects of nature—the sky. He would build astronomical tools unrivaled in their enormity and beauty. He would create scientific tools as works of art and monuments to the human desire to understand. The combination of the scientific and artistic purposes of tools was an inheritance of the Islamic astronomers, whose astronomical tools were never dry and aseptic receptacles of empirical data, but objects of beauty, reflecting in their construction the essential mystery and compelling aesthetic complexity of the sky. To this conception of the artistry of nature, Jai Singh added the Indian concern with mathematics, with patterns and number, and with grandiose cosmology, expressed by his predecessors such as Aryabhata and Bhaskara II.

The result would be perhaps the greatest monuments to human understanding and the desire to learn in existence—gardens of astronomy, drawing one to contemplate the mysteries of life and the universe as easily as they could track the position of the sun or a supernova. The Vedic ritual of ancient astronomy and the distinctly Indian character of astronomy in the subcontinent had one last stand with the efforts of Jai Singh. One of his monuments itself would become witness to the decline and subjugation of a great culture, as the autonomy of the region itself collapsed in the colonial period—a dark era for the subcontinent even as it heralded the dawn of Western power.

Chapter 6
India

Eating the Sun

Perhaps the most startling and amazing phenomena of visible astronomy from earth is the solar eclipse. The bright sun is slowly eaten away by an encroaching crescent, and is finally completely obscured for a brief time, making the whole sky dark, a brief falling of night in the middle of the day. A thin ring of gaseous light hangs around the dark spot in the sky where the sun once was. Not only the sky transforms, but the land reacts as well. Animals make the signs of night, a shadow falls over the land. In the sky one can see the brightest stars, just as in the beginning of night. It never becomes fully dark, as the strange glow of the newly blocked sun clings to existence. For a few minutes, the strange semi-night sets in, and then just as it began, it starts to fade. The arc of light of the sun grows larger and larger, until it once again gains its fullness, and day is restored.

It is one of the most spectacular events one can witness. If one doesn't understand what's going on, it can also be one of the most terrifying. Solar eclipses were seen as ominous events in a number of pre-modern societies, in the East and West. There were also a number of different explanations of eclipses, from the wild to the plausible. The history of explanation of eclipses in India is difficult to get a hold on, but we can make some sense of it. An old myth holds that the sun and moon are both eclipsed by two entities, Rahu and Ketu, which are differently understood in different stories.

Probably the most interesting of the stories holds that Rahu and Ketu are two parts of an originally single individual deity, the asura Svarbhanu. According to the story, the asura was part of a plan to obtain *amrita*, the nectar of immortality for the devas (another group of gods). The asuras agreed to help the devas created the nectar by churning the Ocean of Milk, but the devas intended to have the nectar for themselves and Vishnu formulated a plan to make it so. The churning produced a number of things, and finally the nectar itself. The asuras stole the nectar and wouldn't give it to the devas. Vishnu then incarnated himself and appeared as

A. McLeod, *Astronomy in the Ancient World*,
Historical & Cultural Astronomy, DOI 10.1007/978-3-319-23600-1_6

Mohini, the most beautiful woman in existence, to tempt the asuras and overcome them. She gets them to agree to distribute the nectar equally to the devas and asuras. Of course, since she represents Vishnu, who is in on the devas' plan, the agreement is to allow the devas to drink the nectar first, and the asuras will be allowed none. But Svarbhanu, perhaps suspecting a scheme afoot, disguised himself as a deva, and was offered the nectar by Mohini. As he began to drink the nectar, his true identity was discovered, and Mohini beheaded the asura. Since the nectar had already gone down his throat, though, he did not die. His immortal head became Rahu, and his body Ketu. The head chases the sun, and occasionally catches up to the sun and swallows it. But, because Rahu is just a disembodied head, the sun always emerges from Rahu's throat after he swallows it and flees from him again, after which Rahu once again chases after the sun. This process recurs over and over through time.

A different (perhaps later) account of Rahu and Ketu takes them to be two planets or solid bodies invisible to human eyes, but with the power to obscure the sun and moon (in lunar eclipse). A later and more sophisticated understanding took Rahu and Ketu simply to represent lunar nodes, at which all solar eclipses take place. The knowledge of lunar nodes must have existed in India from a relatively early time, as the mathematical astronomy and the observational abilities of scholars would have made it extremely unlikely they would miss the patterns of eclipse. Most likely, the various explanations relying on Rahu and Ketu were advanced by different people and groups. Just as in our society, experts in a field have a different understanding of it than the general population, and educated laypeople will have a different understanding than the uneducated. Things were likely not very different in ancient India. We might imagine that the story of Rahu the head and Ketu the body was accepted, if by anyone, but the least educated, while astronomical scholars likely understood the lunar nodes and had the more sophisticated understanding. The role of myth in this case is also similar to that of other societies, including our own. Myth is meant to express symbolic and not necessarily literary truths, and likely most people who perpetuated (and still perpetuate) the various stories of Rahu and Ketu do not take them literally, any more than most of us take literally the idea that the north pole is the "top" of the planet or that there is a face represented by the "Man in the Moon".

As with all things, understanding a phenomena can often help to reduce our fear of it. Human psychology is such that perceived danger multiplies and grows more terrifying in the face of the unknown. The growling in the dark becomes to us likely a desperately hungry lion on the prowl, rather than the more likely snoring of a person nearby. Of course, it makes good evolutionary sense that our minds should work this way. Those whose imaginations carried them to the most terrifying scenario in such situations will tend to act more carefully and avoid perceived danger. So in those few times when there *was* actually a lion on the loose, these people survived, whereas their less spooked peers would have been lunch.

The moon was an enormously important celestial object in ancient India. The moon, and the orbit of the moon, has been seen as a sign of fertility by many cultures, including all those covered in this book. The moon has sometimes been depicted as a goddess or female deity, with control over pregnancy and birth. The moon is seen in a number of cultures as having power over times of female fertility. The most likely explanation for this is that a woman's menstrual cycle is generally the same length of time as the synodic period of the moon (the time it takes for the moon to circle the earth and return to the same place in the sky—or alternatively, to undergo a full cycle of phases), 29.5 days. This is a period used to define the month in many calendars through the history of human thought. Our own calendar, though no longer tethered to lunar periods, was also originally based in lunar months. Indeed, the very term 'month' suggests this origin (derived from the older spelling 'moonth'). There was even a belief in some cultures in the West that the full moon had a power to cause drastic character changes or drive one made—hence the lunar association of the English term 'lunatic'. The association of the moon with fertility may have something to do with the superstition common in India and elsewhere in the world up to the current day (including the Maya) that solar eclipses have a negative effect on pregnant women, who must be protected indoors during such an eclipse lest there be dire consequences for themselves and their unborn children. Some Maya take eclipses to be bad omens in general, but especially dangerous for pregnant women, whose children can be killed or born with deformities as a result of the effects of eclipses.[1] Of course, such superstitions are not universally believed in these modern cultures.

The mathematical theory used to calculate eclipses and other astronomical events of regularity in India were outstanding tools in their time, but by the time of Zij astronomy and the attempt to construct tables, people recognized that empirical observation would also be necessary to fully understand the skies. Although there are regularities, these regularities are not as smooth or elegant as mathematical solutions would suggest. Astronomers in India would have to build tools for detailed observation of the skies. Luckily, there had been a deep tradition brought to India by the Islamic world of astronomical observation, built from the cross-section of Arabic, Greek, and Persian astronomy. There had even been Chinese influence, as the connections between the Chinese world and the Middle East grew during the dominance of both regions by the nomadic clans of the Central Asian steppe, most famously the Mongols of Genghis Khan. One of the stories of Indian astronomy, then, is that of the movement from ritual astronomy to theory and mathematics, and finally to observational astronomy. At the same time, Indian astronomy never lost its central characteristics through this entire history. In many ways, the observatories of Jai Singh II stand as the architectural synthesis of the various strands of Indian astronomy.

[1]Milbrath (1999, Chap. 1).

Early Indian Astronomy—Vedic

When the influence of Arabic-based astronomy (itself influenced by Greek and Babylonian astronomy) entered India with the conquests of the 10th–12th centuries CE, India already had its own robust tradition of astronomy. Theirs was an astronomy heavily linked with mathematics, and developments in one area often went hand in hand with developments in the other. The Maya, as we saw in chapter one above, had a very mathematical sense of astronomy, but the astronomers of India, perhaps more than any other until the modern day, were mathematician-astronomers. It is impossible to consider any account of Indian astronomy without also considering the intricacies of Indian mathematics and cosmology. The development of astronomy that led to the magnificent astronomical gardens of Jai Singh was based on a combination of native Indian conceptions of the skies and the mathematical approach to the skies that the ancient astronomer and mathematician Bhaskara II (12th c. CE) referred to as "The calculation of astronomical wonders",[2] and the tools and techniques developed in Arabic astronomy, and which would move into the West as well.

The earliest account of the workings of the heavens and astronomical phenomena that we see in the Indian subcontinent is in the *Vedas*, the collection of ancient texts combining accounts of ritual, cosmology, philosophy, and religion. The Vedas were most likely the result of the synthesis of the culture of the group referred to as the 'Indo-Iranians' or 'Aryans' (from the Sanskrit term '*arya*'='noble', and also related to the Persian term '*Iran*')—the first texts of modern India ("modern" here meaning post-1500 BCE).[3]

The Vedic material is mostly concerned with ritual enactment and verses recounting the glory and in praise of the various gods of the early Vedic pantheon—the gods of both the Persians and the Indians. We find analogous gods in the sacred texts of the early Persians as well, the Avesta. Indeed, the two languages, Vedic Sanskrit and Avestan Persian, are so similar that a person who understands one can also read the other.[4] Astronomy in the Vedas is not central in the way that ritual and hymnal praise are, but it nonetheless makes an appearance. Interestingly, one stark

[2]*Karanakutuhala*, a simplified version of his *Siddhantasiromani*, which was his major work on mathematics and astronomy. For a discussion of Bhaskara's place in Siddhantic astronomy in India, see Rao (2000, 10-12).

[3]While the term 'Indo-Iranian' tends to be used almost exclusively in American and European sources today, likely because of associations with Nazism and racist extremism associated with the word in the West because of the Nazi appropriation of the term as a (historically inaccurate) description of a "white race", in India for the most part the term has no such historical baggage and is still used, just like the revered Indian symbol the *swastika* that was also appropriated by the Nazi's, as a symbol of their fictive "Aryan" heritage. In truth there is little if any connection between the Indo-Iranian/Aryan peoples and the Germanic peoples. If anyone has a claim to be "Aryan", it is Persian Iranians (and just such a claim to identity is made in the name of the nation "Iran") and the people of Afghanistan, Pakistan, and northern India.

[4]Philip Baldi offers an example of these similarities in Baldi (1983, 63).

difference between the early cultures of the Americas and China and that of India is the relative lack of focus on astronomy and the observation and investigation of the heavens. It is not until much later in Indian society that astronomy takes a more central role, and its wonders become the source of attention in society. And even then, fascination with the heavens only inspires a certain small segment of the society.

The sky was a feature of the world that was impossible to ignore, and was featured in important ways in the lives of ancient Indians. There were developed in the Vedic period some of the very same astronomical functions that people elsewhere in the world, including more astronomically obsessed cultures, also developed, such as the astronomical calendar, of various kinds (solar, lunar, and planetary). With this concern, in the Indian case, came a concern with number and ratio, a focus on mathematics that would eclipse that on observation in later Indian history. Just as in Maya astronomy, Indian astronomy was deeply connected to mathematics and numerology. Both cultures held that the numbers involved in divisions of the motions of celestial bodies such as the sun, moon, and planets were ritually and metaphysically significant, and organized important aspects of their cultural lives around these numbers and in ways making these numbers apparent through human action.

The most conspicuous way we see this in ancient India is through Vedic ritual, which was the central and dominant feature of the Vedas. Various aspects of the ritual constructions were related to events in the skies, especially important numbers drawn from a consideration of celestial motions. In Vedic rituals, 360 stones surround the sacrificial area of the fire altar, representing the days in the lunar year, while in the case of the *agnicayana* fire sacrifice, the altar is built with 360 bricks, constructed into the shape of a falcon, to represent the year (Fig. 6.1).[5]

This notion of the 360 day year may sound strange to us, as we know that a full tropical year is about 365 and ¼ days (365.2422 days to be exact). This was known to the ancient Indians as well, but there were, similar to the Mesoamerican calendars, different calendars and counts that gave different values for both years and days.[6] In particular, there were three different calendars or year-counts in use in ancient India. One was what was called the *naksatra*, based on the 27 (later 28) positions of the moon in one orbit around the Earth, the sidereal period of the moon, distributed into 12 months. This leaves a year based on the *naksatra* with 324 days,

[5]Michaels (2004, 249), Kak (1995, 388) maintains that the falcon altar is one version of a more general altar style that could take the form of numerous animals or simply a square, and was composed of five layers and 1000 bricks.

[6]The ancient Babylonians also operated with a 360 day year, and we know that the degrees of a circle, also derived from Babylonian thought, sum to 360. John North (2008, 52) claims that the Indian 360 day year shows that Indian astronomy was influenced by Babylonian astronomy. He says "the figure of 360… is almost as good as a Babylonian signature." This cannot be right, as we have seen that cultures with no contact with Babylon and on the other side of the planet, such as the Maya, nonetheless adopted 360 day year counts. Thus, there is no reason to think that the Indian 360 day year could not be an independent development, as indeed it likely is if it comes from the Vedas.

Fig. 6.1 The "falcon" altar was one variation of an altar used in Vedic fire rituals. 360 stones made up the circle surrounding the altar, and according to some accounts, the falcon structure represented time. The altar was composed of 396 bricks—the 360 of the year as counted in ancient India, plus 36 additional bricks of an intercalary month, or "leap month". This photo is a reproduction of a falcon altar (Photo credit: Arayilpdas at ml.wikipedia)

with 27 days taken as the average number of days of the lunar month. We can see that the *naksatra* year clearly diverges from the solar year, as there are a full 41.25 days fewer in this count than in the solar year count. There were a number of different ways such a calendar was made consistent with the solar year, similarly to the Mesoamerican calendars. The second year type adopted by ancient Indians was the lunar year count, of 360 *tithis*, or 354 days. 354 days is gained from a consideration of the motion of the full cycles of the moon within 12 months, on a count of a *synodic* month, which is 29.53 days (new moon to new moon), rather than the *sidereal* month (return of the moon to a particular spot against the background stars) of around 27 days, on which the *naksatra* is based. $29.5 \times 12 = 354$. Obviously months cannot contain *half-days*, and thus this is averaged out so that 6 months contain 29 days and 6 contain 30 days. Thus, on average, the month contains 29.5 days. This is something that the contemporary Gregorian calendar also involves. You may notice that February (except for leap years) contains 28 days, while 7 of the other months contain 31 days and the 4 remaining contain 30. There are similar sidereal lunar month considerations behind this, even though the months in our calendar have since long ago become disconnected from consideration of lunar cycles, and this is just why it is possible to have multiple full moons within a month, and why the celebrations of festivals based on a lunar calendar, such as the Chinese New Year, the Islamic month of Ramadan, and the Jewish festivals of Passover and Hanukkah, happen at different times within the Gregorian calendar each year (although they always remain within a few months of the Gregorian—such that Ramadan falls in spring, for example, and Passover never in winter).

Tithis, as mentioned above, are day periods that are 1/360th of a lunar year, and thus they do not have the same value as a complete solar day. Since the lunar year contains 354 days, this means that a *tithi* is 122/124ths of a solar day (roughly. Since the lunar day turns out to be not *exactly* 354 days, but rather 354.37 days, there will be a slight error in this determination of the *tithi* as 122/124th of a solar day, but this small error will add to less than a day every five years. Still, something that would likely have been noticed for observant skywatchers).

The third calendar is one with which we will be familiar—the solar year. Ancient Indians understood that the tropical year falls between 365 and 366 days, and based the solar year on this number. Our own solution to the problem of the fractional value of the solar year in days is to add a leap year day every four years, thus accounting for the 0.25 additional days past 365 accrued every solar year. The Mesoamerican calendars, as we saw in the first chapter, did not add such considerations, and so their seasons slowly slipped out of sync with the calendar. This was a conscious decision on the part of the Maya, who were aware of the nature of the solar year.

We can see in the example of the number 360 that was so significant to Vedic ritual, then, that the greatest concern was with the various lunar-determined calendars, such as the *naksatra* and the lunar year. Part of the reason for this concern may have been the more apparent location of the moon along the background of the stars, and also the organization of seasons and periods easily with lunar phases and positions. Phases and positions of the moon are one of the most easily accessible calendars, far more apparent to all than the motions of the sun through the zodiac. And a close and lengthy observation of the motions of the moon will eventually tip one off to the secrets behind one of the most startling celestial phenomena witnessed—the *eclipse* (solar and lunar).

The motions of the moon are also thought to be linked to another major feature of Vedic Indian society—the drink *soma*. Soma was a concoction including a drug unknown to us but believed to be either a hallucinogenic or perhaps an amphetamine, which was consumed in ritual context, as a method for opening communication with divine realms beyond the human.[7] This use of sacred drugs for shamanic connection with the divine is not unique to Indian culture. Almost every civilization has a similar feature built into its ritual and religious structure. The Maya kings and nobles, for example, through their bloodletting sacrifices, entered into trances in which they were thought to be in contact with the underworld of *Xibalba*, which exists parallel to (perhaps "behind" our own world). It may have been the blood loss involved in such ceremonies that created or contributed to this trancelike state, which was recognized and revered as of mystical importance. The Huichol people of western central Mexico have a practice of shamanic ingestion of

---

[7]This view of *soma* is controversial, with some scholars arguing for a more benign identification of the drink. A brief overview of the debate can be found in Jamison and Brereton (2014, 30-32).

hallucinogens (particularly peyote), which serve a similar purpose of opening up the portal to a spirit realm, in which mystical knowledge inaccessible to humanity in its normal state can be revealed. And this kind of thing is not limited to Non-Western cultures. Consider the practice of the Catholic Church of eucharist, still in existence today. The body and blood of Christ is represented by (or rather *is identical to*, after the process of transubstantiation) unleavened bread and wine. Why wine? The mind-altering effects of alcohol, especially in the absence of food (the traditional practice is to observe some period of fasting before receiving the eucharist) will inevitably create a sense of connection to the divine and the wider cosmic mysteries. Islamic society, which shunned alcohol, used coffee (and its stimulant drug, caffeine) in this way—indeed it was the Islamic world that introduced the rest of the globe to the drink that has become a staple of many cultures including our own in the West.

As *soma* was identified with the moon in the Vedas, its ritual consumption was supposed to correspond to the monthly period of the moon. The psychological and social effects of the moon, or at least the culturally perceived effects of such, have been part of the folk wisdom of many cultures, including our own. Almost everyone has heard the myth that a full moon brings violence, madness, and criminality to the fore in people. Indeed, this is the origin of our English word 'lunatic'. The moon's revolution and phase shifting was thought to bring about recurrent psychological changes. The moon has also been linked to the menstrual cycle[8] (and there may have been an even closer tie with lunar phases in the premodern world in which people watched the moon and organized their lives around its phases). The moon was also linked to the determination of seasons, of which the ancient Indian calendars marked *six* (of two months each) rather than the more standard four or two, based on the solstices andequinoxes. Those familiar with the climate of most of India (all of it below the Himalaya range) will understand the reasoning behind this. The six seasons were, in the following order: spring, summer, rain, autumn, winter, freeze. We see here the standard four seasons, with the addition of the rain season and freeze. In mid to late summer, most of the subcontinent experiences the Monsoon, which is a period of humidity and precipitation that lasts over a month. The rains produced in this season are strong and dense, far moreso than most of what we experience in the United States, for example. The rest of the year in the region is relatively dry, and most of the water that sustains life falls during this season.

Which brings us to the freeze season. Winter in the subcontinent (except for mountainous areas in the far north) is far milder than anything we experience in most of the United States. Only for part of December and early January do temperatures ever reach near freezing at their lowest extent during the night in north. The months surrounding this are then the freeze months—during which the temperature may fall low enough to cause frost and kill certain plants. We see, then, that in Indian culture, climate was used alongside of astronomical motion to

---

[8]Law (1986)

determine the seasons. While the astronomical motions, particularly that of the sun, are of course linked to these climate features (the sun figures as a cause on some level of explanation of both the rains and the freeze), they are not directly linked in the way that the solstices and equinoxes are. And this may in part explain why the people of ancient India do not seem to have been as concerned with observational astronomy as those of Mesoamerica and China. It would have been seen as less critical to agriculture, and thus survival, than it was in China and even Mesoamerica where the solar seasons closely mirrored the rainy season for much of the region.

The connection between the configuration of celestial objects and the seasons may have also been part of the reason behind the ceremonial and ritualistic character of Vedic astronomy. According to the early thinkers, to be an astronomer and to be a Vedic priest were one in the same, as the knowledge gained from astronomy was necessary for proper performance of Vedic ritual. A passage from the *Yajurveda* section on astronomy (*jyotisha*) explains:

> The Vedas have, indeed, been revealed for the sake of performances of sacrifices. But the sacrifices are laid down on the sequence of appropriate times. Hence, he who knows astronomy, which is the science of specifying time, knows the sacrifices (and, so, is a Vedist). (Subbarayappa and Sarma 1985, 1)

The ritualistic character of Indian astronomy was to remain a major feature throughout much of its history, and in the final section of the chapter we will return to look at the implications of this for Indian astronomical thought in general, and later developments in it.

From very early on in Indian astronomy, we can see that mathematics and the mathematical understanding of the sky becomes central. Perhaps more than any other culture, including the Maya, ancient Indians were numerologists, and it often seems that astronomical observation serves as a means for understanding mathematics in the early period than the other way around. Ultimately for the Maya, as well as for us in the modern world, astronomical mathematics is a means, a tool used to understand the sky, which is our focus. For people in ancient India, it seems to have been the reverse. It was mathematical discovery and innovation that was the goal, and astronomy the means. This is the reason one often hears major figures like Aryabhata, Bhaskara I and II, and others referred to as astronomers and mathematicians. The emphasis should probably be on the latter. These men were every bit as much, if not more, mathematicians than they were astronomers.

There is no question that observational astronomy took a back seat to mathematical theory in India through much of its history, perhaps until the Islamic period. We might ask why this should be the case, given that astronomy so clearly involves viewing the heavens, and massive amounts of data concerning the locations and motions of celestial bodies. The answer to this question, I think, goes far beyond consideration of astronomy in particular to some foundational views underlying much of early Indian intellectual culture.

In the famous text *Bhagavad Gita*, a segment of the much larger epic *Mahabharata*, concerning fundamental philosophical and religious issues facing humankind, such as commitment, duty, life and death, and the ultimate purpose of human life, a key distinction is made that illuminates an important view in ancient Indian thought (that comes up at various points in Western thought as well throughout its history). Krishna, the charioteer who represents the ultimate spirit, or God, is speaking to the warrior Arjuna on the field of battle. One of the many things Krishna tells Arjuna in his extended teaching that comprises the *Gita* is that material causation is determined. There is nothing Arjuna or anyone else can do about the eventual outcomes of events in the world. A certain battle or war will be lost (or won), a ship will sink at sea, you will walk into a certain restaurant at 5:32 PM exactly. All of these things are determined. Krishna explains that he, the divine principle, makes these things the case, and that there is nothing anyone can do to make them otherwise.[9] The motions of the planets and the stars are a good demonstration of this. They go about their way, unalterable in their paths, constantly moving in predictable and mechanistic cycles (even if it took many of thousands of years for humanity to discover the principles behind those regular motions).

While this is the case, it should not lead us to become "fatalistic", in that we simply decide there is nothing we can control or change and cease acting altogether. Although we cannot change anything in the deterministic physical world, there is one thing that we can control, and this thing is ultimately far more important, and even more *real*, than the physical world—the spirit, mind, or soul. While it may be the case that the outcomes of our lives are fated—we will get a certain job, live in a certain place, injure a certain body part, etc., our mental state is not determined. We can become attached to the external world and thus suffer, because this world is impermanent as it is always changing, or we can instead seek to understand the true nature of being, the *atman* (or self), which turns out to be identical to the universal or divine spirit itself (Brahman), and thereby be both released from suffering through attachment, and also have a better understanding of the world created by the desires of the spirit.

The ancient Indian conception of reality might seem to us to be the *reverse* of that often assumed in the West. We often take the physical to be foundational, with any mental properties or substances (if such exist at all, which some thoroughgoing physicalists actually reject!) arising somehow as a function of or otherwise from the physical. This is why many today see the empirical sciences as "real" knowledge, and the human arts and sciences as somehow make-believe, secondary, or otherwise irrelevant. One passage from the *Rigveda*, the most philosophically-oriented of the Vedas, explains the arising of the world thus:

> All that existed then was void and formless: by the great power of Warmth was born that One. Thereafter rose *desire* in the beginning, desire, the primal seed and germ of spirit. *Rigveda* 10.129 3-4 (Griffith 1896)

We see here that mental states are seen as primordial and foundational to the world. It was not that mind arose from the physical world, but rather the physical

---

[9]*Bhagavad Gita* 18:61 makes this point.

world arose from mind. Not *human* minds, of course—the *Rigveda* is not claiming that humans were the first things to exist in the universe. But the mental—mental states, thought, spirit—these things existed in the beginning, and everything that arose came from them. And indeed, how do we know this is not the case? It is a far more plausible position with much more evidence than we often give it credit for being. All will admit that our experience of the world, including our sensation (which we take to be the hallmark of empirical science), is a function of our minds. There is nothing we see, feel, or do that is not "in the mind". When we see something, our vision is not of the world, but is an image created by the mind—by the firing of neurons receiving information and creating images. This is the case for all senses and all thoughts. Not only do we not sense the world "as it is in itself" (even if the world is directly responsible for our sensations), but we can never get outside of our heads, if you will. This is why there is always a question about whether our senses are accurate, and why we cannot sense the thoughts of another, for example. Since this is the case and the mental is far more fundamental to our experience, why not consider the possibility that it is fundamental for being in general? This is just what many ancient Indian thinkers held.

It is for this reason, among others, that some have argued that Indian observational astronomy may not be native, but instead adopted from earlier Babylonian and Greek forms. This view holds that the Indians took the observational data and the methods of skywatching from these cultures. Perhaps even the mathematical concern was not developed in India, but was itself borrowed from these other civilizations, even though astronomical mathematics were certainly developed in India equal to or beyond that of any other cultural area. There is still controversy surrounding the issue of the extent of originality of native Indian astronomy. Some claim that almost all of Indian astronomy is derived from that of the Persians, Greeks, and Arabs, while others argue that much of Indian astronomy is unique and that where India adopted the astronomical techniques of other cultures, they assimilated them to the existing structures of Indian thought.[10] Certainly there is no question that Indian astronomy throughout its existence, like Indian thought more broadly (religious, philosophical, scientific, etc.) was highly syncretistic, adopting, adapting, and building upon knowledge gained from numerous areas of the globe. One of the key features of Indian thought in general has been its ability to integrate and synthesize seemingly disparate strains of thought. This ability is no less in evidence in the realm of astronomy than it is in the more familiar realms of philosophy or religion.

There are good reasons to reject the view that Indian astronomy is derivative. The earliest texts suggest that astronomers in India were taking their own observational data, and the mathematics developing around this astronomy was likely an Indian creation. This does not mean that there was no outside influence on Indian astronomy. We can in one important sense see a kind of external influence on

[10]North (2008, 171-175) is an example of the "Indian astronomy as derivative" view, while (Rao 2010) takes the opposite position.

Indian astronomy and that of the entire wider region of southern and western Eurasia. One interesting mathematical relationship that both Indian and Greek astronomer/mathematicians seem to have discovered or inherited was the view that the distance from the Earth to the Sun was about 500 times the length of the diameter of the Earth. Through modifications to this view as a result of further development, the Indian and Greek estimates began to diverge by the time of Aryabhata (6th c. CE) and Ptolemy (1st c. CE). Some have argued that Aryabhata's estimates were based on those of Ptolemy, but this seems highly unlikely, as Aryabhata derives the formula from other information, and never mentions Ptolemy in his work (which he would have had no reason to do). Although there was almost certainly not direct influence by Ptolemy, could there have been indirect influence? Perhaps, but there seems to me an even more likely explanation.

As we see in Aryabhata's work, he derives the estimate of the sun's distance based on facts about a number of other distances and diameters, such as that of the moon. In one clever move he determines that the distance of both the sun and the moon from the earth are about 100 times their own diameters, a fact that is likely supposed to explain the similarity of the apparent sizes of the solar and lunar discs, which accounts for the possibility of totality in a solar eclipse, a strange and interesting coincidence of nature. Once he has this information, he can derive the diameter of the sun and moon, and their distance to the earth in terms of earth diameter, which he determines to be slightly less than 500. Many of these assumptions about the Sun's distance in diameter seem to be assumed in both Aryabhata and Ptolemy, and this suggests that the source of this view is earlier than both of them. There is a perfectly reasonable explanation for this, and it can be found in philosophy, language, and culture. It is a fact that scholars sometimes forget.

The ancient Indian philosophers, in their theories of the person and the world, as well as in their debate methods and assumptions, closely resemble the ancient Greek philosophers, and reading works of the two traditions together will inevitably make one wonder whether there was cross-influence between the two. Indeed, there are certain early texts from both cultures that mention the other, and engagements with the other. In one unjustly neglected philosophical text, the *Questions of King Milinda* (*Milindapanha*), we are even presented with a philosophical debate between a Buddhist monk and a Greco-Bactrian king. There are even more startling similarities in language between Indian languages and Greek and Latin-based languages. The Sanskrit term for fire is 'agni', while the Latin term is 'ignis', which is the root of the English word 'ignite'. The Russian term is even closer, being identical to the Sanskrit, 'agni'. Likewise, the Sanskrit term to breathe, atman, later signifying the soul or self, has its exact mirror in the German term for the same, atmen. The similarities between Indian and Persian languages and culture has already been mentioned above. Such amazing coincidences led scholars to investigate, and eventually discover, that the explanation for the similarities between European and Indian languages (and cultural features) is that they share a common origin. Indian, European, and Persian languages are all developments from a common ancestor language known today as "Proto Indo-European". And where there is a shared language, there is generally shared culture as well. Scholars believe that the source of Proto Indo-Europeans was a cultural group and people who occupied the Eastern European steppe. For some reason, either due to migration, conquest, or the widening of their cultural sphere, the PIE peoples extended their range east into Asia and west into Europe. The eastern wing became the people we know as the Indo-Iranians (or 'Aryans'), the cultural ancestors of both the Persians (Iranians) and the Indians, and the western wing became the Greeks, Romans, and other Europeans. The PIE languages and cultures retained a remarkable resiliency even through their transmission to different peoples and cultures encountered on the movement of PIE peoples and culture, and this likely means that the PIE languages and cultures became dominant in the areas they entered, whether due to conquest or other means.

Not only aspects of language, but also key aspects of culture followed the PIE peoples. The centrality of the horse, the wheel, and the chariot, for example, seem to have been a feature of PIE cultures.[11] Many of us associate chariots with ancient Rome, the chariot races being a well known aspect of that culture, and made even more visible to contemporary people through popular Roman-themed Hollywood films like Ben Hur or more recent ones like Gladiator. This place of the chariot in Roman culture was an inheritance from its PIE ancestry, as it was for the people of Greece, Persia, and India.

If cultural features as well as linguistic ones could be transmitted through PIE peoples, it should come as no surprise to us that astronomical knowledge could also be thus transmitted. We can see in cases such as that of the Americas that astronomical knowledge is among the most basic and central knowledge of many peoples, and is transmitted as part of one's earliest human inheritance (although things are much different in the modern world in which we no longer depend on the sky). There are deep similarities in the astronomical systems and the star lore of peoples throughout the Americas who as far as we know never directly contacted one another, and which cannot be chalked up to human similarity in development and worldview, as they are unique to the Americas. This may be an inheritance of the common origins of Native American people in the complex of peoples who migrated into the Americas from Asia tens of thousands of years ago, just like the centrality of the horse, wheel, and chariot throughout European, Middle Eastern, and Indian culture is a remnant of the shared PIE culture that spread into these areas. The equation of the distance of the sun with around 500 times the Earth's diameter may be such a shared inheritance of the PIE peoples. Certainly these peoples had some astronomical knowledge and beliefs, and these would likely have been fairly central to their lives on the steppe, helping to determine the calendar and track the seasons, which would have been more critical for them in the harsher climate they inhabited.

It is indeed safe to say that Indian astronomy is unique, as much as Greek or European astronomy can be. All of these systems share a family resemblance of necessity—not because one or some of them influenced the others, but because they share a very basic cultural template, independently of the Babylonian influence.

---

[11]David Anthony discusses these features of PIE culture in Anthony (2010).

The pre-Siddhantic astronomy of the Vedas and other cultural material was the "native" astronomy of India, which was reinvigorated after 500 CE with the introduction of new astronomical ideas from outside of the subcontinent, from Arab and Greek astronomers. The Siddhantic period, which is the term for this renaissance made possible by the synthesis of new ideas, saw an explosion in both concern for and innovation in astronomical and mathematical thought. The Indians were not unique in having their astronomical (or any other) renaissance inaugurated by an influx of knowledge from outside. Pretty much every intellectual awakening in the history of humanity has happened in this way—but not all traditions have admitted it. The case most well known to most of us is that of the European Renaissance of the 15th–17th centuries, which followed a smaller one in the 12th century. The former, more famous renaissance, had a number of external sources. First, European intellectuals discovered (or in some cases rediscovered) the texts and artworks of the ancient Greeks, which had been unavailable in Europe for most of the Middle Ages (even thinkers like Aquinas, obsessed with Aristotle, had little of the Greek material), because the Greek area of influence was controlled by hostile Islamic states—most importantly the Ottoman Empire, which was based right in the heart of the Greek world, at Istanbul (which had formerly been Constantinople, the city of Constantine during the late and Eastern Roman empire, and a Greek city before this). Second, beginning in the late 15th century, new ideas from parts of the world such as the Americas, East Asia, and Africa entered into Europe with the beginning of oceanic exploration and "discovery" by various European nations. This source of new ideas spurring the Renaissance is not generally admitted or credited, as the European narratives preferred to attribute their intellectual awakening to the ancient Greeks, who they saw (incorrectly) as their European ancestors—likely in part due to the influence of bold syncretistic thinkers like Thomas Aquinas and the later scholastics, who had been so successful in their integration of Aristotelianism that they managed to make something that had once been seen as foreign, exotic, and irrelevant (as Greek thought did to most Europeans at the time—indeed the whole idea of Christendom, on which the very idea of Europe is based, did not extend to the Greeks, but stopped with Western Europe and the western church) appear to later Europeans in the Renaissance as not only familiar but part of their own cultural inheritance. Greece had become "Europe". The debt to the rest of the world often still goes unacknowledged. The Europeans encountered the grand civilizations of Mesoamerica in the late 16th century, with all the ideas considered in chapter one and many more, and first seriously encountered the Celestial Empire of the Ming Dynasty in China in the 14th century (the exploits and almost completely fanciful stories of Marco Polo from the 13th century aside). It is no mere coincidence that the dates of the Great Renaissance in Europe just happen to overlap with the dates of contact with these distant lands, their people, and most importantly their culture and ideas.[12]

[12]Walter Mignolo argues that some of the ideas and language of this "renaissance" itself has colonial motivations (Mignolo 1995).

And of course renaissances work this way. New and innovative ways of thinking do not happen spontaneously from within—they have to be spurred by variations in the environment. Intellectual change mirrors biological change in this (although not in every) way. A creature that is adapted for a specific environment does not all of a sudden just develop a change—this is not how evolution works. Change in a creature's evolution is either precipitated by changes in environment, requiring development in order to adapt to that new environment, or by mutations that give members of the species in question an advantage in their existing environment over those without the mutation, which is then perpetuated to future generations. And just as the European Renaissance flourished based on the influx of ideas from around the world, the Indian astronomical renaissance started during the introduction of unique Arab and Greek ideas about the sky into the subcontinent, around the 6th–10th centuries CE, with the rise of the Islamic states which eventually moved into the subcontinent as well. While there had been contacts between Indians, Arabs, and Greeks for a long time before this, the cultural influence of Arab thought become more pronounced during the conquests and migrations of Islamic regimes into India. The Indian renaissance that followed the synthesis of these new ideas was not limited to astronomy. We see a number of innovations in art, literature, science, religion, and culture. The religious tradition of Sufism arose through a synthesis of Islamic ideas in an area heavily influenced by Buddhism and Hinduism in what is today Afghanistan.

However this is not the whole story. Astronomical thought was beginning to come into its own even before the new ideas that would spark the renaissance gained wide currency in the region. The work of the 6th c. CE astronomer and mathematician Aryabhata, along with that of the 7th c. commentator Bhaskara (a mathematician in his own right) transformed astronomical knowledge in the region, and represented astounding advances in both mathematics (especially) and astronomical understanding. It is in the work of Aryabhata and his commentators that we see the special and unique concern for mathematics in astronomy emphasized and carried to a degree beyond what we see in the astronomical systems of most other cultures.

Siddhantic Astronomy

Siddhantic astronomy gets its name from the so-called *Siddhantas*, astronomical texts influenced by Greek and Persian thought, combined with traditional Vedic astronomy. Characteristically of most of Indian astronomy throughout history (except for perhaps the very earliest times), the nature of Siddhantic astronomy was highly mathematical. In this period and for many years afterward, astronomy and mathematics were not seen as separate pursuits. Just as astronomy and philosophy were closely intertwined in the Chinese world, astronomy and mathematics were indistinguishable in India. We might wonder then, given the highly mathematical nature of our own contemporary astronomy, to what extent the Siddhantic astronomy of India was similar to that of astronomy as it is practiced today. There are of course a number of similarities, but the

Siddhantic astronomy of India was different in that it was *driven* by mathematics, rather than empirical observation. While modern astronomy uses mathematics as a language for understanding phenomena, its basis is in empirical data, and mathematics are used to formulate theory. Siddhantic astronomy was different—mathematical theory often served as the basis, making astronomy more of an a priori (or theoretical) pursuit than an empirical one. Truths about the world and the universe could be discovered through understanding mathematical truths. Mathematics was a model of the world. This much modern astronomy seems to have adopted from India. But for the ancient Indians, mathematics went beyond just being a language through which to understand the world. It illustrated the deepest features of the world—thus making discoveries in mathematical theory (which can be done without access to empirical data at all) was to make discoveries about the world. There was a necessary connection between mathematical theory and the world that allowed discoveries in the latter to be made by investigating the former.

The Indian astronomers, in the Siddhantic period and before, were also concerned with questions of cosmology, as far back as the Vedic period. As in other cultures, we see that the conception of the human and human meaning in ancient India was connected to that of the cosmos. And uderstanding the role astronomy played in the Indian conception of humanity may help us understand why someone like Jai Singh II would have thought that having a massive observatory in his backyard would somehow cement his place in human history (as indeed it probably will).

Indian Cosmology

Although, like every other people with access to the sky, the ancient Indians observed and tried to make sense of the events of the sky, what kind of cosmology did they have? What significance did they see in the celestial phenomena, and how did this mesh with their worldview in general? Unusual events such as eclipses, comets, and supernovae—cultures generally have some explanation for such things. And the human mind hardly ever rests content with just noting the existence of some phenomena—we attempt to understand it as consistent with some overarching system that explains the world around us, and gives it meaning and significance to our own lives. Contrary to some of the claims one will hear in modern science, humanity is not now, and has never been, concerned completely with *how things are* independent of us. Indeed, certain features of quantum mechanics seems to show us that we never can view nature from a purely "pristine" perspective, without including ourselves in the picture. Every perspective from which we view nature is a perspective of *ours*, and is fundamentally human, whether we like it or not. The ancients understood this, even while we today try to ignore it or act as if we have transcended the merely human standpoint. The observation and understanding of nature is at the same time an observation and understanding of *ourselves*. The human is inextricably linked to our conception of the world and how it works. There is no "view from nowhere", which is the ideal of the modern sciences.

Perhaps no philosophical tradition understood this fundamental truth better than that of ancient India. From the earliest philosophical texts, the *Upanishads*, forward, into famous texts like the *Bhagavad Gita* (part of the epic *Mahabharata*) and even in some sense into the offshoot tradition of Buddhism, there is a view that the cosmos itself, or what is most central to the identity of the cosmos, *Brahman*, the "infinite spirit" is none other than the most inner self of each individual (*atman*—"spirit" or "soul", derived from the term for "breath"). Understanding the cosmos, they held, is first and foremost a matter of understanding the true nature of the self, as the two are non-different. There have been many ways of understanding or making sense of this basic claim throughout the history of Indian philosophy, but it is a major feature of the tradition that is shared by the majority of the schools and theorists. The Indian tradition in general seems to find intuitively correct the basic position that the self and the cosmos are not distinct entities, that the two are inextricably linked, and that indeed the individual self is a microcosm mirroring or even identical to that of the whole universe. In this, the ancient Indians were somewhat close to both the Native American civilizations and the Chinese, who also saw a key continuity between the human and the celestial worlds. Perhaps only in the Indian case does this rise to the level of identity.

Cosmological thought in India was meant to make sense of the key features of the natural and celestial changes, including the standards for time, in terms of human creation and sustenance, similarly to robustly developed cosmologies elsewhere in the world, such as China, Egypt, Mesoamerica, Persia, and Greece. Literate cultures such as these left us texts recounting the detail of their cosmologies, and their meanings (although in the case of Mesoamerica, since much of their textual culture was destroyed by the Spanish, we have to rely on inscriptions from sites and artifacts or post- conquest accounts such as that of the *K'iche'* Maya *Popol Vuh*). The most fundamental aspect of Indian cosmology, another similarity to the Maya perhaps, is the focus on time. The Indian system of time is vast and enormous, like that of the Maya, and is organized around certain metaphysical principles that draw our attention to connections between the tenor of an age and human characteristics. According to traditional chronology, the fundamental unit of time is the "day of Brahma". Time is thus fixed to the god of creation, whose life supports that of the cosmos. As one might guess, a day of Brahma is almost inconceivably large from the human perspective, and in our terms such a time period stretches 8.64 billion years. For some modern perspective, this is roughly twice the age of the Earth. One day of Brahma, while it seems amazingly large to us, passes the same interval in the life or Brahma that a solar day does for a human. In this cosmic version of humanity writ large, Brahma's lifetime mirrors our own. Thus, Brahma's lifetime is 100 Brahma years (or one *Maha Kalpa*), or 8.64 billion times 360 (the number of days in the precorrected solar year or the lunar year based on *tithis*. As our current universe is the same age as its creator, Brahma, existing only insofar as Brahma exists, this means that the lifetime of the universe is 311.04 *trillion* years.

While the philosophical motivation for and significance of the Maya calendar is not as well known, due to the loss of classic texts, that of the traditional Indian calendar is fairly well understood. This raises another interesting point and a distinction between the Mesoamerican and South Asian cases. The Long Count calendar of Mesoamerica and its associated cosmology and mythology was not only ritually significant and explanatory in a cosmological sense, but was also pragmatic, based in concrete present time for those who used it. When the Maya spoke of years in distant *baktuns*, well before the current, they did so from a discrete and definite position in the calendar itself, which was used throughout the Classic Period to mark important dates, such as the accession of rulers or the conquest of cities. (Indeed, the use of the Long Count in this way is one of the features determinative of the Classic Period). In ancient India, although a cosmology was developed that included the counts of many thousands of years and yugas, kalpas and maha-kalpas, there was no practical calendar that took these particular dates into account, anything like the Maya Long Count. So for all intents and purposes in the Indian calendar, these vast dates were more theoretical than practical. One never knew *exactly* where one was within a yuga or any other large chunk of time. All parties agreed that we inhabited (and still do) what they called the *Kali Yuga* (age of Kali), on which more below, but there was a great deal of disagreement as to when the *Kali Yuga* had begun, how long a *yuga* lasts, and how many other *yuga* had passed since the beginning of this universe.[13] This alone shows that there could not have been a calendar based on these counts in use, as such issues would have had to have been worked out satisfactorily before the adoption of a calendar that would be understood by a wide segment of the population. Not *every* calendric problem must be worked out before a calendar can be used practically, but at least a few must be, including most basically agreement about the units of time and (roughly) when they apply, how long they last, etc.

Consideration of the concept of the *kali yuga* will show us that the "calendar" in question here was more religious and ideological in its basis than practical. It is not that the people of ancient India were simply poor calendar makers or didn't possess the skill to calculate the ages as efficiently as the Maya. From an investigation of their astronomical mathematics, the Indians certainly did have the necessary knowledge and ability to create such a calendar. The existence, or lack thereof, of various cultural, philosophical, scientific, or other concepts or features in a culture most often does not have to do with intrinsic abilities or knowledge (after all, as much as some like to think differently, humans are roughly of the same intellectual ability everywhere they exist, and have access to roughly the same information), but rather has to do with the cultural focus and interest. We can learn a great deal about a culture through looking at its literature, philosophy, science, and religion. That is, we can see what were the greatest concerns of the society in question. The fact that the Indian " calendar" does not and cannot actually function as a practical calendar does not show us that Indians did not know how to make calendars, but rather that they were not concerned with keeping time in the precise and detailed way the Maya were, for example. Such a conception of time had no import for the people in ancient India. One reason for this may be revealed in the philosophical background.

[13]There were similar disagreements between sources in the Classic and Postclassic Periods in the Maya world on formative dates in the calendar, corresponding to local differences.

In the Indian case, we see a focus on religious and ritual events rather than historical events such as political events throughout the history of the subcontinent. It is for this reason the dating of particular events that we know took place in early Indian history is on shakier ground than that of the other two cultures we have looked at, the Maya and the Chinese, both of whom were highly concerned with meticulous date keeping. If we think about the Maya and Chinese cases, one thing we notice is the importance of rulership as an organizational feature of culture, and the tying of precision concerning dates and creation of history as connected to rulership and the organization of the state. Both Maya and Chinese society developed in such a way that rulership and governing was in part determined by the control of both the calendar and the written records. Histories were produced by imperial courts or states in early China, for example. Also in China, rulers named periods of time after segments of their own rule, so that their identity would be built into the determination of time itself. The ruler became the essential man in part because the ruler became identified with time, with the continued motion and sustenance of the world. This was true in both the Maya and the Chinese cases, but perhaps no civilization can be said to have been more concerned with time than the Maya. In ancient India, it was not command of time that was seen as essential to power or rulership, but command of *ritual*.[14] Ritual played an enormous role in Indian society, as a dominant feature of the Vedic culture inherited from the PIE peoples. The key to efficacy in general, whether controlling the outcome of battle or attaining a thriving life, was ritual knowledge. Because of this, rulers in India tended to see themselves and present themselves as guardians of ritual, or, as Indian thought developed and the notion of religious law or teaching (*dharma*) superseded that of ritual, guardians of *dharma*. The role of time in such a construct was minimal, as ritual and *dharma* are thought to transcend time. Thus the ruler controls not time, but something far more important, more essential.

Given this focus, we can see why it would be relatively unimportant to keep a detailed and precise calendar based on the multiple units of time of the traditional cosmology. An understanding that there *were* these periods, and that we currently inhabit a particular place in this larger calendar, was enough. Which brings us back to the issue of the *Kali Yuga*. We have been in the Kali Yuga for all of recorded history, according to the many thinkers who have dealt with the issue. A *yuga*, or "age", is variously determined by different scholars, but all are in agreement that the current age is that of Kali. The features of the *yugas* form a degenerative mythology, similar to that seen in a number of other traditions, in which humans descend from a kind of state of perfection in the beginning to a more lowly and humble state in the present. Such mythologies are often tied to morality, attempting to explain seemingly ineliminable negative features of human life such as mortality, suffering, and lack of power to control our environment by attributing to humanity some kind of critical neglect. In the story of the Western theistic religions, it was the curiosity and disobedience of the first people, their eating of the apple of knowledge

[14]Thapar (1997, 120-130)

forbidden to them by God, that led to the downfall of humanity. This story is still one of gradual decline. Notice that the biblical story of the human fall from grace is not one of *immediate* and total degradation. The first people presumably can live forever, and with the first sin comes mortality. But the direct progeny of these first people don't live the standard 80 or so years—they live for many hundreds of years. Their children live a bit less long, and their grandchildren even less long, until eventually after many generations human life has become so short that we cannot even, on average, last a full century. The effect on mortality of the original sin were not immediate in the story, but manifested itself over time. The Indian story of the *yugas* is similar. In the first, *satya* (truth) *yuga*, there was high morality, people lived indefinitely, everyone and everything was beautiful, and humans were able to control the world around them. It was a time of paradise. In succeeding *yuga*, all of these things went into decline, such that the second, *treta yuga* saw people who were less morally good, less beautiful and powerful, and less long-lived. There was a further decline in each *yuga* down to the *Kali yuga*, which is the final age in the cycle of *mahayuga*, after which there will be a new creation and the cycle will start over. The *Kali yuga*, then, is the age in which humans are as degraded, morally bankrupt, powerless, and fragile as it is possible for them to be. It is the "end time" in terms of human life and culture (although not in terms of Brahma, which creates anew at the end of each *mahayuga*. It is quite appropriate, then, to name the age after Kali, who is the goddess of destruction and thus renewal.

It seems to be a universal feature of humans to see one's own age as fundamentally more difficult, more corrupted, and more frightening than those of the past. Perhaps part of the reason for this is that the past is set in stone—all of its outcomes are known, and from this perspective it appears safe. We know how the story ends, how the actions of people or the random impositions of nature affected the lives of those in the past. Another possible reason for the feeling that the present is more difficult is the visceral closeness with which people inhabiting the present feel the problems presented in the present. The struggles we face, and the suffering we endure, are direct and personal, and leave us with a sense that they must be more profound than those of prior generations.

The universe itself that this time directs, organizes, and influences is similarly vast, inhabited by innumerable gods and superhuman beings as well as animals and humans. The scientifically known universe, to observers in Ancient India, was of course the same one known to all other early observers—the Earth, the Moon, the Sun, the visible stars, the planets, and the mysterious "astronomical wonders" that occasionally seemed to shatter the regularity of the cosmic motions—supernovae, comets, and eclipses. Although it was not until relatively late in human history, and in the West, that the nature of the first two of these phenomena were properly (we assume) understood, we will see in this chapter that Indian astronomers were intent on, and to a large extent were successful in, determining the regular patterns of eclipse of the sun and moon. Cracking the "code", discovering that (especially solar) eclipses do follow a complicated but true pattern, requires feats of both observational and mathematical genius. Luckily, this is something that astronomers in India had plenty of. And, as mentioned before, in the Siddhantic period they had the aid of an influx of new ideas that, synthesized with native astronomical and mathematical forms, revolutionized the understanding and appreciation for the heavens in Indian history.

Islamic Thought and Astronomy, and Its Influence on India

In the early 7th century of the Common Era, hundreds of years after the declining Roman Empire in the west had been invaded and sacked numerous times then brought to its inglorious end, a new empire began to take shape on the sleepy western shores of the then relatively unimportant Arabian peninsula. The land of Arabia was almost total desert, inhabited by families of tribes and roaming nomadic clans, or "Bedouins", a social phenomenon suited to and quite necessary for survival in such a harsh environment. It was nowhere near the center of civilization. The people of the more developed and cosmopolitan regions of what we today know as the Middle East, Persia in the northeast, Syria and Mesopotamia in the northwest, with their thousands of years of culture and history, saw the Arabian peninsula as a kind of "backwater", populated with uncouth nomadic barbarians. The Arabs were to the Persians and Syrians what the Germanic tribes were to the Western Romans. But as was the case with Rome and the Germanic tribes, the mighty were destined to be brought low, and the conquerers would become the conquered.

In the first years of the 7th century, a profound and powerful new religious and political movement developed in the Hijaz region of Arabia. A 40 year old man from the Banu Hashim clan of the Quraysh tribe in the city of Makkah (Mecca) claimed to have received a startling and definitive revelation from God, transmitted through the medium of the angel Jibril (Gabriel). This man, now prophet, Muhammad ibn Abdullah, would bring the word and the rule of God to this formerly ignorant people (the Arabs), and use his message of transformation to turn Arab society from what it was into a pious, God-fearing, organized, unified, committed, and inspired force. With his religious and political message, Muhammad and his early followers managed to, albeit with many years of conflict and hardship, reshape the raw, disorganized, and fractured chaos that was Arab society, in which most energy was expended on fighting between clans and tribes, into a powerful and faithful community. This community would eventually direct its attention outward rather than inward on itself, and would construct one of the greatest empires the world has ever known. Although this empire began more as a political construction (it was certainly seen this way by its earliest leaders), as it grew it became solidified and unified through religious commitment, and with this glue the Islamic movement would go on to have a far greater influence than even the early Islamic empires ever did. Even non-Arabs, and those who never fell under the dominance of Arab rulers, would eventually become entranced by the allure of Islam, adopt many aspects of Islamic culture, and would carve out their own empires every bit as impressive and opulent as those of the early Arab rulers. Some of these, of course, developed in the Indian subcontinent, and it was here that the astronomical and mathematical learning that had been developed via the contact between peoples afforded by the earlier Arabic and Islamic conquests was brought to India and fused with Indian thought—and to dramatic effect.

The key tenets of Islam, the ideological basis on which Muhammad created his community, are today well known. The term '*islam*' or "submission" (to God) entails that the members of the Islamic community are to become bound by their shared obedience to the law of God, as revealed by Muhammad, and thus to become a kind of new "chosen people", similar to the Hebrews before them. There is much continuity with Judaism and Islam in this respect, that they are both law and community based approaches to religion in which the political aspect is *part* of religion. Indeed, Islam was greatly influenced by Judaism and Christianity, and the written revelations to Muhammad, today known as the *Qur'an* ("Recitation"), accept the authenticity of the revelations to the Jews and Christians represented by the Hebrew Bible and the Christian Gospels. Islam sees itself as the continuation and completion of the message of God that was earlier revealed by previous prophets, including the Jewish prophets and the Christian prophet, Jesus (Isa).

This community started in the early 7th century by Muhammad was united mainly by shared belief in the message and based on the leadership and strength of personality of Muhammad himself. The beginnings of the Islamic movement were inauspicious. In the early days of Muhammad's movement, he and his followers were driven from his hometown of Mecca by an antagonistic clan, and were forced to resettle in the city of Medina. It was in Medina that Muhammad's movement gained strength sufficient to stand on its own and became a powerful force to be reckoned with in the Hijaz, and later throughout the Arabian peninsula and even the wider world. This forced exile from Mecca and exodus to Medina is commemorated by Muslims in their observance of fasting during the lunar month of Ramadan, which is the month according to which this "new exodus", or *Hijra*, took place. Muslims also date the beginnings of the Islamic calendar from the *Hijra*, with the first year designated as that beginning after the move to Medina. As of this writing, 1435 years have passed since Muhammad's followers migrated to Medina. The Islamic calendar year is 622 years behind the Gregorian calendar we use today in most of the world, because this calendar is based (roughly) on the birth of Jesus. We can determine from this that the *Hijra* took place in 622 CE[15] (according to the Gregorian calendar).

---

[15]CE, or "common era" and BCE or "before common era" are the most common neutral terms in use to designate years in the Gregorian calendar among historians, supplanting the older AD, or *Anno Domini* ("Year of the Lord") and BC ("Before Christ"). While AD and BC do make clear the origins of the Gregorian calendar in the Christian chronology of the life of Jesus, the adoption of the terms CE and BCE are meant to secularize the accepted calendar. Yet another proof that our designations of time do have a special meaning and significance. In a pluralistic, secular, and increasingly globalized world, many of us desire to use temporal references that reflect our ideals, not those of one particular group within society. CE and BCE are meant to make the designation of time relevant and unoffending to non-Christian groups, including those of other religions as well as the non-religious. We cannot change the *basis* of the Gregorian calendar, which has been in such wide use for so long and is now a global standard, without massive difficulty, but we can certainly change our names and designations for these standards.

The Islamic calendar, like some we have already encountered in this and earlier chapters, is a lunar calendar, but unlike some of those we have seen, it is based on the first appearance of the crescent moon as the beginning of each month, rather than the new moon. This significance of the crescent in the lunar calendar likely contributed to the centrality of the image of the crescent in the Islamic religion. The crescent and star symbol often represents Islam, which in many of its guises forbids depiction of individuals (as steps toward idolatry). Atop a holy building such as a temple or *mazjid* (mosque) one often finds a crescent, identifying the building as an Islamic place of worship, much as crosses perched atop buildings identify Christian churches.

As the Islamic movement spread (both religious and political—the two were always intertwined in Islam just as in Judaism, both religions of the *law*), Arab rulers and elites came into contact with a number of different civilizations and traditions. As always when there is a meeting and synthesis of cultures, there was an intellectual renaissance in Islamic lands in the years after the conquests. The (at first) strictly Arab religious and political movement spread through conquest into North Africa, throughout the lands of the Persians, into India, and west into Europe, through the Balkans and Greece, at the doorstep of western Europe on its eastern side, and even into western Europe from the south, as the Muslim conquerors moved into Spain and established *al Andalus*—a place that would become one of the centers of Islamic learning in the early medieval age and the cradle of some of the Islamic tradition's greatest thinkers.

In this explosive expansion into the region much of which had once been part of the Roman empire, but had since developed in its own fragmented ways, the Arabs both brought their culture and knowledge to these lands, and also assimilated knowledge from the lands they conquered. In the early years of Islamic rule, the Arabs neither sought nor allowed conversion to Islam by their subjects, likely in part due to the desire to differentiate themselves as an elite class among the people of their new realms. Regardless of how hard people will try to put up barriers around themselves to keep out the cultural influence of those around them, culture always finds a way. Once there is contact between peoples, the dissemination of culture both ways simply cannot be stopped. Eventually, the strictly Arabic nature of the Islamic movement gave way, and other peoples began to join and influence the direction of the culture in its various branches (although Islam would continue, until this day in many sects, to have a very Arabic-based slant—the holy language is Arabic, in which the word of God was uniquely revealed, and the cultural trappings of the Arabs were fairly clearly codified in much Islamic norms. The difference was now that non-Arabs could become like Arabs too).

One form of learning that flourished from fairly early on in the Islamic empire was science, including astronomy. Although in pre-imperial Arabic thought there had not been a great deal of attention to science, partly due to the condition of the peoples of the peninsula at the time, who were mostly poor Bedouins or relatively provincial clans, the renaissance spurred by the integration of so many new lands, peoples, and ideas galvanized the culture, which became one of the foremost scientific traditions in the world. Some attribute the birth of science to Western Europe or further back to Ancient Greece, but the true birthplace of modern science was in the lands of Islam after the conquests. Perhaps the birth of science here was linked to the combination of the knowledge and methods of the various lands the Arabs conquered with the strongly systematizing insistence on unity of the Islamic religious tradition itself.[16] Since the beginning, the Arab-Islamic tradition had been concerned with *oneness*, the fundamental principle at the core of the religion itself. The principle of *tawhid*, the ineliminable oneness of God, was the banner around which the religion itself formed. Muhammad's movement was in some ways a reaction against both the polytheistic religion(s) dominant in the Arabian peninsula before his time, and to perceived innovations in monotheistic religions, such as Christianity, that appeared to him to deny the fundamental oneness of God. Perhaps more than any other tradition, Islam took this unity as its core.

Perhaps this insistence on oneness and unity had a role to play in the origins of science in the Islamic world. The ideology of the scientific enterprise, after all, depends on this key assumption of the unity and singular explanation of the seemingly multiple phenomena of the world. The principle of "reductionism" has always been a key aspect of science, for better or worse. To give a scientific explanation is not *just* to give a mechanistic or materialistic explanation for observed phenomena, as some people today assume. Perhaps the central aspect of it is the reductionistic. A scientific explanation takes seeming multiple explanations and reduces them to a *single* explanation (wherever possible—and it is not always possible!). Science is uncomfortable with (or intolerant of) the existence of more than one explanation for the same event, or what is often called *overdetermination*. Instead, it demands fundamental explanations to which all others can and must be reduced. This fundamental assumption underlying science has caused some unease at times through the history of science, and thoughtful people have sometimes worked hard to try to offer a justification for it that goes beyond simple intuition or choice. Probably the most famous attempt is that of William of Ockham, whose principle now referred to as "Ockham's Razor" states that one should posit the fewest possible entities necessary to explain a phenomenon. Simplicity is to be preferred. This, however, is more a slogan than a justification. Why should we prefer the simpler explanation, or the single rather than the multiple? It cannot be predictive power—we can have equally predictive theories that posit many more entities or involve more layers of explanation. Perhaps the dirty little secret of science is that the scientific method itself cannot be established by scientific method. And this is fine, but ultimately we have to admit we can't have "turtles all the way down"—as much as modern humans might want it to be the case, the universe does not privilege our scientific understanding or dub it as necessarily true.

---

[16]Seyyed Hossein Nasr points out that there is in fact deep Indian (and Persian) influence in early Islamic astronomy. He writes: "In astronomy, the Muslims continued the tradition of Ptolemy, while making extensive use of the knowledge of the Persians and Indians. The first astronomers in Islam, who flourished during the second half of the second[Islamic]/eighth[Gregorian] century in Baghdad, based their astronomical works essentially on Persian and Indian astronomical tables." (Nasr 1987, 168).

For all of its usefulness, science is a human invention and tool as is religion or any other aspect of culture. It is useful to remember this, especially when we consider the power and value of ancient astronomy. This is not to say that science does nothing *different* than religion, art, and other human constructions. But science is a human interpretation of the workings of the world. There is no such thing in the world independent of humans as the "Theory of Relativity", even though this theory may be an accurate way of understanding actual features of the world. One sometimes hears it said that the world is constituted by mathematics—but this has things in reverse. Mathematics is how *we* understand the world, not how the world operates in itself.

This is also not to say that science has been represented in the Western context, while religion dominates Non-Western cultures. Every culture I know of (and likely every one in human history), has had both science and religion. This is certainly true of the cultures discussed in this book. The history of science, like the history of astronomy, is not a story limited to Europe and the Middle East.

The science of the Islamic world was precipitated by the earlier developments of the Persians and Greeks, whose lands were subsumed in the Islamic conquest. There was an ancient and established practice of studying the heavens and the natural world in general in both of these cultures, and the introduction to the Arabs of the knowledge and techniques the Greeks and Persians had developed spurred development of their own science, particularly mathematics and astronomy, in which the Arabs became major innovators.

The introduction of Islamic astronomy into India happened with the movement of Islam into the subcontinent, which happened in a number of waves. In the earliest years of the burgeoning Islamic empire, the frontiers were in the difficult mountainous regions bounding Persia to the east, which was the transition point of Indian culture, in current-day Afghanistan. While today we think of Afghanistan as the "middle-east", in the ancient world this region was closely tied to the culture of the Indo- Europeans of south Asia. It was a center of the Buddhist movement and the birthplace of Mahayana Buddhism. The Gandhara region, and its situation at the confluence of Greek, Persian, and Indian cultures, produced some of the richest philosophical, religious, and artistic innovations of the ancient world. The region is well known for the so-called "Gandhara style" Buddhist sculpture of Greek and Indian influence. In this region East and West came into contact and engaged with one another. The previously mentioned text *Questions of King Milinda* originated in this region. The text dramatizes a philosophical debate between Milinda (Menander), a Greco- Bactrian king, and Nagasena, a Buddhist monk, concerning the central metaphysical and ethical questions of Buddhism. The ancient culture of the region is probably better known today for the unfortunate reason that it has come under assault by the militant fundamentalists who have asserted their control of the region in recent years. The ancient and majestic Buddhist earthen statues at Bamiyan were infamously destroyed by the Taliban in 2001.

Although early in the 8th century there were contacts between India and the Islamic world in this region, mainly in modern-day Afghanistan and western Pakistan, which also had noticeable effect on Indian astronomy and mathematics, the Arabs were unable to conquer the subcontinent early on, as their forces were defeated by Hindu confederations against the already overstretched invaders. It was not until the beginning of the 2nd millennium, in the 11th–13th centuries, that Islamic regimes were able to get a foothold in India and rise to dominance, and by this time it was no longer Arabs but Afghan Muslims who fought to conquer the region. The entire region eventually came under Muslim control, and the expansive Delhi Sultanate existed from foundation in 1206 to its fall to Babur, in the early 16th century, a Turkic (Central Asian) Muslim invader and descendant of the famous conqueror Timur (Timur the Lame—*Timur-i-leng* in Persian, the "Tamberlaine" of Marlowe's play), and also more distantly of Genghis Khan, who established the Mughal dynasty that ruled most of India until the time of the English.

The Persian-Islamic astronomy that entered into contact with Indian astronomy at the time led to the development of what some call "*Zij* astronomy",[17] which was concerned with the construction of astronomical tables, and therefore also with increasingly sophisticated observational instruments, with which to make more and more accurate tables. Just as the European observer Tycho Brahe discovered, in order to make accurate astronomical tables, one needed precise tools. In Europe and India both at the time of the beginning of these respective projects, the precision that could be had with the available tools was far too low. We might well ask—what difference does this make? While knowledge is certainly useful in itself, when we consider the role accurate astronomical data plays in understanding our universe, we see just how important it is.

Although the mathematical innovations in astronomy of Islamic astronomers had for the most part been anticipated by Indian astronomy, one area in which Islamic astronomy aided that of India was in observational technology. Despite the brilliant mathematical methods that throve in India from the beginning of the Siddhantic period of astronomy, Indian astronomy clearly lagged behind in the area of observation. Comparatively rudimentary observational tools were in use in India before the Islamic period. With the entrance of Islam and rise of Zij astronomy, Indian astronomers adopted and came to use tools such as the astrolabe, armillary sphere, and the equatorium—a device for determining planetary positions invented in Islamic Spain.[18]

The spherical trigonometry that made possible devices like the astrolabe and equatorium was well understood by Indian astronomers, who immediately appreciated their significance and used them (and similar tools) to give greater precision to their observations of the skies. In China, these tools, though known, did not come into wide use, although the armillary sphere remained an important tool there, in the land of its highest development, down to the modern period. In India, use of the Islamic astronomical tools would lead to development in the construction of

---

[17]See Kochhar and Narlikar (1995, Chap. 1).

[18]The *solar* equatorium was known and discussed as far back as the 5th century CE, but the planetary equatorium first enters history in *al Andalus*.

instruments that would come to a pinnacle in the work of Jai Singh II in the 17th century, whose magnificent work still stands today as a testament not only to the science of observational astronomy, but also to its artistry and humanism. The latter are two areas of astronomical thought we have all but discarded today in our fervor for materialistic science. In the Islamic astronomical tradition, we see a commitment to astronomical tools as themselves objects of beauty and aids to contemplation of nature. Astronomical observation in the ancient and medieval world was not the reductionistic search for data and numbers gained by use of aseptic and lifeless tools that it has largely become today. Astronomy was a human pursuit, and the design of astronomical tools was not only a matter of generating precision (although it was that too), but also one of humanizing the material. In no tradition was this more true than in the Islamic astronomical tradition. The attention to artistry in this tradition led to some of the most beautiful astronomical tools in human history—tools not solely meant for the production of astronomical tables, but whose use would lead to aesthetic and philosophical contemplation of the universe being studied. These tools reminded their users of both the meaning and significance of the projects they engaged in, and also served to humanize the process. What astronomers did could not be divorced from humanity and all of our concerns, our emotions, our aesthetic sensibilities, our hopes and fears, and our rationality, without becoming some kind of monstrous abomination. To observe and attempt to understand the skies, the Islamic astronomers thought, was to peer into the structure of God's creation. This was necessarily more than just numbers and charts, but held the key to the *spiritual* significance of creation. Astronomy, to the Islamic-Indian thinkers, just as it was to the Mesoamericans and the Chinese, was not just a science, it was a core area of human contemplation. The attempt to understand the skies could not be distinguished from the project of determining the meaning and significance of our own lives, such that to engage in astronomy was in some sense one of the most important philosophical pursuits. Reflecting on the meaning of life was reflection on the heavens.

In some ways, contemporary astronomy retains the traces of this kind of conception. When we watch television shows on cosmology in physics, for example, there tends to be a solemn or sacred tone to the proceedings, with majestic music playing reminiscent of medieval Masses. Statements are delivered about the big bang and other such phenomena in powerful and authoritative baritone with the cadence of priests. Carl Sagan was one of the progenitors of this. Watch an old episode of his "Cosmos", and it's often difficult to tell the difference between Sagan and an Anglican priest declaring the Creed. Popular depiction aside, the professional practice of astronomy rarely has the kind of humanistic contemplation these general presentations sometimes approach, and certainly nothing like the ancients or medievals. Indeed, contemporary science tends to shun such things as "unscientific" or standing in the way of understanding, rather than contributing to it. Our notion of science tends to be reductionistic, and anything beyond the most pared-down material explanation is rejected as "spooky" or irrational. I believe this aspect of modern science is a flaw, that comes from the domination of science to the neglect and detriment of all other human arts. Just as in the world of humans, when one entity has hegemony over everything else, that which makes the world vibrant and meaningful dies away. We cannot rely on science as the end-all and be-all of human life any more than we can rely on one method to handle every problem in the world, or one useful tool such as a fork to help us eat everything we eat (we'll have problems with pudding or juice!). History is awash with examples of peoples who insist "only this", and the ends of such myopia should serve as a sobering warning to all of us.

The magnificent scientific and artistic achievement of Jai Singh might be seen as the end and the pinnacle of Islamic influenced Indian astronomy.

Return to Jai Singh

Our skip forward to a very different period, at the beginning of the modern world, may strike some as strange, given the focus of this book on ancient astronomy. However, the work of Jai Singh II can reveal a great deal about early and indigenous Indian ways of thinking about the cosmos and of engaging in astronomy. Jai Singh was in many ways an antiquarian, and was interested in reviving neglected or long forgotten aspects of ancient Indian culture, including Vedic ritual and ancient methods of investigating the skies and charting coordinates of celestial objects. He was also a gifted scholar and had enormous knowledge of both Eastern and Western astronomy.[19] Jai Singh was responsible for some of the most artistically magnificent (and astronomically accurate) instruments in human history, in his five major observatories, built at his own city of Jaipur, as well as the Imperial Mughal capital of Delhi (today known as "New Delhi", the capital of the Republic of India), Mathura, Varanasi, and Ujjain. Jai Singh's creations and his attempt to restore the greatness of Indian culture and learning through a focus on traditional ritual and astronomy came during a particularly dark period in Indian history. Jai Singh's story has much to tell us about the will of the human spirit to promote intellectual ideals and learning, even in the face of the precipitous decline of society around them. Interestingly enough, many of the greatest and most creative innovations in human history came about during periods of strife and difficulty.

Jai Singh II was born into royalty, an heir to the kingdom of his great-grandfather Jai Singh I, ruler of the Rajput kingdom of Amber, a Rajput kingdom of Rajasthan (in current day northwest India) centered on the city of Amber. Jai Singh II would go on to build a magnificent new city close by, named after himself and still known today by this name, Jaipur ("City of Jai [Singh]"). It was in this city that he built his most impressive observatory. His father, Bishan Singh, and great-grandfather Ram Singh I, son of Jai Singh I, were vassals of the Mughal empire. The Rajput kingdoms in general lost their autonomy when they came under the control the Mughals, to whom they for all intents and purposes

[19] Hooja (2006, 671)

became vassal kings. The Mughal empire had precipitous beginnings in the subcontinent. It was inaugurated by the central Asian conqueror Babur, who according to the traditional account certainly had a lineage that would suggest imperial glory. He was a descendent of the world-conqueror Timur (better known in the West as Tamerlane, or "Tamberlaine" of Christopher Marlowe's play) on his father's side, and the most famous warlord of history, Genghis Khan, on his mother's side. Although Babur's homeland was in the central Asian steppe (as his famous ancestors before him) he moved into the region around Kabul in modern day Afghanistan, and it is there that he is buried, despite his forays into the south and eventual conquest of much of the Indian subcontinent.

Babur's empire became known as the Mughal Empire, and his descendants controlled the subcontinent almost completely for hundreds of years, from its establishment in the early 16th century until the early 18th century, though it existed in some more or less weakened form all the way to the 19th century, when it was finally brought to an end with the capture of the final "Emperor" Bahadur Shah II in Delhi after the famous Rebellion of 1857 against British rule. After Babur's time, his descendants consolidated their control over the subcontinent, fighting battles against independent kingdoms throughout the region, and exerting indirect political control over others. The Rajput kingdoms were among those under the thumb of the Mughals. Intrigue in Jai Singh's own hereditary house was caused by the orders and influence of the Mughal emperors.

In Jai Singh's day in the mid to late 17th century, there were strong connections between the family and the Mughals, as Jai Singh acted in military service of the Mughals, and was duly rewarded for this. His successors were less successful. Jai Singh's son Ram Singh was less influential with the Mughals and the prestige of the family suffered. By the time the next successor to the kingdom of Amber was established, Ram Singh's grandson Bishan Singh, the family's fortunes had hit bottom. Bishan Singh acceded to Amber kingship without official Mughal recognition, which was unprecedented for the family. The emperor Aurangzeb engaged Bishan Singh in near-impossible military tasks, and the future for the family looked bleak when his son Jai Singh II acceded to the throne at 11 years old on the final day of 1699. When Jai Singh came to power, the family was near penniless. The Mughal emperor Aurangzeb seemed to have no particular love for the ruling family of Amber or the Rajputs in general, and this antipathy continued during Jai Singh's rule. Luckily for Jai Singh, the later years of his rule coincided with the beginning of the decline of the Mughal empire, and through a combination of political and military acumen, Jai Singh was able to expand his direct power and influence. By 1727, his resources were vast enough to allow him to construct the magnificent city of Jaipur, where he was able to take shelter from the crises buffeting the Mughals in Delhi, who were suffering military losses to both internal and external forces. The most ultimately successful of these forces was to finally take control of much of the subcontinent by the middle of the century. In the Battle of Plassey in Bengal of 1757, the Englishman Robert Clive and his compatriots dealt a fatal blow to the Mughal-linked Nawab of Bengal. It was this that established British rule in India, which was expanded for the next hundred years, during rule by the British East

India Company, and then by rule of the British Crown (known as the British Raj, or "rule") after the 1857 rebellion. This rule lasted until 1947, with the establishment of India and Pakistan.

Jai Singh II ruled his relatively small kingdom during the twilight years of Mughal power in the early 18th century. Jai Singh was not only interested in increasing the power of his rule and his kingdom, but he also had intellectual interests, in Vedic ritual, religion, and astronomy. Jai Singh, and the Rajputs generally, were Hindu, while the Mughal overlords were Muslims. The idea of Hinduism as a unified religion did not exist before the Muslim entrance to the subcontinent. 'Hinduism' of course has a similar etymology to 'India' itself, both being based on the Indus (Sindhu) river, which occupies present day Pakistan. The Muslim travelers (first) and conquerors (later) classified the variety of religious practices and expressions in the subcontinent together as Hinduism—the religion practiced below the Indus. While this was always a somewhat artificial construction, the Muslims were right to notice important similarities between the varieties of religious expression in the subcontinent. Most of them had origins in the Vedas of some kind or another, and almost all of them (with the exception of Buddhism and Jainism, which had by this time almost died out in India) accepted the authority of the Vedas.

Nonetheless, Vedic ritual had declined in India in part due to the domination of the subcontinent by the Muslim Mughal Empire. While some early Mughal rulers were famous for their tolerance, pluralism, and love of learning (the emperor Akbar chief among them), after Akbar's time, beginning with his grandson Shah Jahan (the emperor responsible for the building of the Taj Mahal), more fundamentalist and intolerant rule became the norm, and Hinduism and its practices, including Vedic ritual and astronomy, went into decline.

While Jai Singh II is most famous for his impressive observatories, which I discuss below, a lesser known fact about him as that he also worked to revive a number of Vedic rituals.[20] This fact should not be seen as independent of his astronomical thought or the purpose of his observatories. Indeed, understanding Jai Singh's concern with ritual may help us to answer some important questions concerning his astronomy. Specifically, why did Jai Singh fail to use modern tools from the West in his observations, to compile his astronomical tables? Why did he rely on seemingly outdated large masonry astronomical tools? We know that he had knowledge of the telescope, for example, as a telescope is included among Jai Singh's collection of astronomical tools.[21] And he likely had an understanding of how the telescope worked and what could be done with it, as he also had access to

[20]There are numerous accounts of Jai Singh's performance of Vedic rituals long abandoned. Gupta and Bakshi (2008, 117) collects these accounts, and Jai Singh's religious practices and views in general.

[21]Rima Hooja (2006, 671-672) discusses Jai Singh's envoys to Europe to bring back astronomical tables, tools, and methods from the West, and shows that he was well-versed in European astronomical thought.

numerous astronomical texts from the East and West. Yet he decided to forego the use of these tools.

The massive observatories Jai Singh built were used to create catalogues and the observations there resulted in the text *Zij-i-Muhammad-Shahi*, a Zij astronomical table which, as its title suggests, was dedicated to the Mughal emperor Muhammad Shah, whom Jai Singh wanted to honor as his patron (remember that for all of Jai Singh's concern with ritual and learning, throughout this time one of this major goals was the improvement of his family's fortunes). Following other *zij* texts, Jai Singh's text included data on planetary tables on motions and locations, star charts, and other astronomical data gained through observations at his observatories as well as existing astronomical texts. Some have claimed that Jai Singh's *zij* was derivative and ultimately unimportant as an astronomical text because it had been surpassed by texts in Europe.[22] I think this is to neglect Jai Singh's purpose in creating the *zij* as well as his reasons for constructing his observatories. It was not that Jai Singh did not understand what the Western technology could do or that he didn't know about the astronomical tables of the West—rather he likely had different purposes in mind for which the Western material and tools may not have been suited. At least charity would suggest that this is the answer, unless we find reason to believe otherwise.

The observatories of Jai Singh, his most lasting legacy and most impressive construction (still visited today by tourists to the cities in which they still stand), were also his masterpieces. It is not just in their artistry and scale that they are significant (though this is what they are generally appreciated for today), but also for the uniqueness and importance of the *method* of astronomical observation they made necessary. As mentioned above, Jai Singh attempted to revive various Vedic rituals in his kingdom. He was the last known individual to perform one of the most well-known Vedic rituals of the ancient world, immortalized in the Indian epics Mahabharata and Ramayana, the "horse sacrifice", which involved (among other things) the releasing of a ritually purified horse to roam and the sacrifice of this horse on its return home. Jai Singh also revived practice of another quintessential Vedic ritual, the *Yajna* ritual, involving incantations and offerings being entered into fire at the altar. Jai Singh's insistence on performing these Vedic practices even in a neglectful era and also in the face of the Mughal antagonism toward such practice shows his commitment to Vedic culture and thought, and there is no reason to think that this concern did not extend to his astronomical thought.

The observatories themselves were conceived and constructed on a massive scale. The largest of them, Jai Singh's own Jaipur observatory, takes up the space of an entire park, with a number of large stationary masonry instruments that range from the smaller ones of about the size of a car to the massive and towering instruments like the *Samrat Yantra* ("Supreme Instrument") that reaches almost 100

---

[22]Andreas Volwahsen (2001, 103) writes: "[Jai Singh's] main work, *Zij Muhammad Shahi*, the tables dedicated to the Mogul emperor, are essentially based on Ulugh Beg's tables and on European sources."

feet high and takes up a large section of the observatory. The Samrat Yantra is by far the most eyecatching and impressive of the instruments at Jai Singh's observatory from afar, but the traveler soon discovers a wealth of other instruments with as much detail and artistry, if not enormous scale. Even the smallest instrument in the observatories, however, is massive by the standards of the time, and even our own time. The Samrat Yantra rivals the largest optical telescopes in the world in size, such as the Gran Telescopio Canarias in Spain and the Keck telescopes in Hawaii. Of course, the *Samrat* is not a telescope, but has a different purpose.

The majority of Jai Singh's instruments were designed to enable the charting of coordinates of celestial objects, including the sun, moon, planets, and stars. A number of the instruments have duplicate purposes—for example, some of the functions that can be realized by the *Samrat* can also be performed on other *yantras*. There are some *yantras* with one narrow specific purpose, and other *yantras* (with the Samrat as the primary example) with a variety of purposes.[23] Some of the instruments in Jai Singh's observatories are meant for imaging of objects like the sun and moon. For the most part, his observatories were meant to be tools enabling him to construct the most accurate astronomical tables possible. Or at least this was *one* of the purposes of the observatories. Jai Singh claims that the *Zij-i-Muhammad-Shahi*, and by implication the observatory he built to help compile the data contained in it, was created in order to improve on the available astronomical tables, which clearly included errors. A more accurate mapping of the sky required more accurate observations than had previously been possible, and therefore better observational instruments. The problem with astronomical tables had always been that, given the limits of the instruments observers worked with, the coordinates of objects would not be exact. So instruments capable of greater precision were needed. One way of gaining greater precision in measuring movements of the sky is to measure them with *larger instruments*. Or at least so Jai Singh reasoned.

Why is this? If we think back to the example of the gnomon, discussed in Chaps. 2 and 4 above, we can begin to see why this is. The smaller the gnomon, the smaller the shadow that will be cast by it. If we have a shadow cast by a gnomon of one inch at its longest morning extent at sunrise on an equinox day, then the furthest extent of travel between the end of the shadow at sunrise (to the west) and the end of the shadow at sunset (to the east) will be two inches. So an entire day of 12 h is represented by two inches of shadow. Given such a small space, our hour lines will not be very accurate. It will be hard to distinguish 11:30 am from noon, let alone distinguish 11:55 am from noon, which will be impossible. The difference of the length of the shadow cast by the gnomon at these times will simply be too small for us to observe. Thus, the smaller the gnomon, the less fine-grained our determination of time can be. On the other hand, if we have an enormous gnomon (say the size of the *Samrat*'s gnomon edge), and the maximum length of travel is 200 feet over the equinox day, then more fine-grained determinations of hour angle are possible. The

[23]The most complete description of the operation of the various tools of Jai Singh's observatories is given in Sharma (1995).

difference between 11:30 am and noon will be a matter of *feet* rather than tiny fractions of an inch. The shadow will also move much faster than that cast by the small gnomon, as it has a lot more space to cover through the 12 h of the day. The larger the gnomon, the longer the shadow, the faster it will move, and the more fine-grained our distinctions of time units based on it can be (of course there are other problems with the gnomon-sundial discussed above).

Just as this is the case with determining time based on gnomon shadows, this also works for determining coordinates of celestial objects on the celestial sphere (based on right ascension and declination). The larger an instrument is, the more fine-grained our determination of its coordinates can be. And thus, if we are looking for more accurate and precise astronomical tables, the development of instruments allowing for greater precision in marking coordinates is of vital importance.

But, as we saw in the chapter on Chinese astronomy, bigger (by itself) is not necessarily better. While in principle a shadow from a larger gnomon can give one a more accurate reading, it is also the case that the larger the instrument, the more vaguely defined the shadow cast by it will be. That is, there will be a region from the beginning of the shadow to the full darkness, and this will tend to negate any possible added precision to be gained by the larger size of the instrument. That is, unless one has a shadow definer, as did Guo Shoujing in 13th century China.

But greater precision and accuracy in observation this was not Jai Singh's only purpose in building his observatories. After all, if it was only more accurate astronomical tables he was after, there was no need to resort to redundancy in building his tools. The *Samrat yantra* and perhaps a couple of other instruments connected to it would have been sufficient to serve his purposes for constructing the *Zij*. Jai Singh must have known this, for he had a deep understanding of the way his instruments worked and the principles on which they worked. To think that he would have been unaware of redundancies in his collection given this would be absurd. When one visits the observatories, and knowing Jai Singh's positions on the Vedic ritual, as well as reflecting on the methods of compiling data using these instruments, it is hard to avoid a particular conclusion.

One of the most striking features of the observatories, even today in their present state of relative disrepair (they would have been kept in much better shape in Jai Singh's time, but they've not been used since shortly after his death), one can clearly see the geometric artistry with which Jai Singh designed the observatories as wholes and each of the instruments within them. The observatories are all in rectangular park areas, with the instruments laid out with respect to one another in ways that complement their geometric order. They are not simply randomly placed or placed in ways one may think each instrument would be maximally operational. Rather, when one walks to the top of a *Samrat Yantra* (as one used to be able to the first time I visited the observatory in Delhi, but sadly public access up the staircase to the top is no longer permitted), one can see the aesthetically chosen geometric order of these instruments. It looks as much like a church or a painting as it does an observatory. Clearly Jai Singh intended to build not only an operational observatory, but also an artistic monument to the intellectual understanding of the cosmos, one that featured in its very structure the order and beauty perceived in the cosmos itself, in part through the use of these instruments.

Jai Singh's largest observatories are those in Jaipur and Delhi, and the only currently operational observatory is that in Jaipur (the Delhi observatory was badly damaged in conflicts throughout the later centuries, especially during the 1857 rebellion). All of the observatories are laid out in similar ways. One of the key questions concerning the observatories has to do with their function. While some have pointed out that the instruments Jai Singh were less accurate than those used in other parts of the world such as in Europe, this consideration becomes a bit baffling when we discover that Jai Singh had knowledge of European tools such as the telescope. There are a couple of other mysterious facts that complicate things further. One is that Jai Singh's instruments were redundant. Instruments with the same exact functions would be placed in the same observatory (there are two versions of the *Samrat Yantra* in the Jaipur observatory, for example). In addition, there are some instruments that not only are inoperable, but could never have been functional. All of this is strange if Jai Singh's sole or main purpose was to develop more accurate astronomical tables.

Of course, there need not be any deep or intellectually stimulating reason for these facts. It could just be that Jai Singh desired to build observatories to demonstrate his wealth and power, and that it didn't matter to him whether they were operational simply because they were intended to be primarily decoration, or to signal to the world Jai Singh's majesty. Perhaps he was trying to recall the period of Akbar, presenting himself in the same "enlightened ruler" vein. But this explanation is not completely satisfying. Although all rulers are certainly concerned with their images and projecting power (this was not something new to Jai Singh), none of them build magnificent observatories like this, and it's unclear that constructing such observatories would have lent to any sense of power or wealth for Jai Singh at all, more than the scale and grandeur of his city of Jaipur would have gained him. The observatories would be a single stone on a robe encrusted with diamonds. Its inclusion or exclusion would not have made a difference. So it is most likely there are other reasons for the observatories.

A consideration of the function of the tools of his observatories may help us here. And as the Jaipur observatory is both the largest and contains the most instruments, it should be taken as the comprehensive observatory of Jai Singh. The others are more or less smaller-scale versions of the Jaipur observatory.[24]

The most impressive, and multi-functional of Jai Singh's instruments is the *Samrat Yantra*, of which there are two at Jaipur—a smaller and a larger. The gnomon of the large Samrat rises nearly 100 feet. At basic, the Samrat Yantra is made up of two main sections. The gnomon section is built as a right-angled triangle—the end of it rising straight into the sky perpendicular to the ground, and

[24]Although they are not *replicas* of Jaipur. The *Zij-i-Muhammad-Shahi* notes that the four observatories at Jaipur, Varanasi, Ujjain, and Mathura were built to check the data based on observations at the Delhi observatory. Thus, the other four observatories, including Jaipur, were more likely copies of the Delhi observatory.

with a ramp rising from the south side of the gnomon tip to the top of the gnomon, at the angle of elevation of the pole. Thus, the rampway points at the pole star, and ends at the top of the gnomon. The angle of elevation of the Samrat Yantras will of course change with location, as they are fixed and dependent on the latitude of the location in question. We know that the pole will be found at the same elevation angle from the horizon as one's latitude angle above the equator. This will make it the case that different Samrats built in different cities will have different elevation angles in the ramps. Jaipur, for example, sits at 26.93 degrees north, while Delhi is at 28.36 degrees north. A very significant difference when it comes to astronomical observation! The elevation angles of the Samrat gnomons in these two locations then were built for their location—with Jaipur's being about 27 degrees. Notice that this issue also shows part of the difficulty with portable astronomical instruments—there has to be a way to accurately adjust them to local elevation (and other conditions), which becomes harder and harder to accomplish with large instruments that allow for greater precision (Fig. 6.2).

There are stairs all the way up the ramp to the gnomon on the Samrat Yantra, so that it's possible to move up and down it in order to make observations. These stairs come into use mainly for nighttime observations. The situation in the day is much easier. The second main section of the Samrat is a structure built parallel to the celestial equator, curving up at the sides, following the arc of the celestial equator. The inside edges of this ring are flat, so as to allow for observations of hour angles in the daytime, and on the larger instruments, there are also stairs on these "quadrants". The whole thing together, then, looks like (and is indeed laid out much like) an elevated sundial, with the gnomon pointed to the pole and the dial in the plane of the celestial equator. The Samrat Yantra, although it can play the role of a sundial as well, has some functions a sundial does not (a sundial in principle could do these things as well, but due to the way it is constructed, it would be difficult if not impossible to make certain observations using it).

For daytime observation, use of the Samrat is straightforward, and it works in basically the same way as a sundial. The gnomon edge will cast a shadow onto one or or the other quadrant, and, given the large size of the gnomon, one will be able to make a detailed reading of the time (given calibration using the equation of time, as discussed above). To determine the elevation of the sun, the stairs on the gnomon ramp can be used. One simply takes a pole up the stairs until they reach the point at which the shadow cast by the pole first appears on a quadrant. It is the angle represented by this point on the stairs that corresponds with the elevation of the sun, with the values represented in Fig. 6.3.

For nighttime observation, the process is somewhat more complicated, since the stars and planets don't emit enough light for visible shadows to be cast. Thus, two people are needed for determining coordinates with the Samrat. For hour angle (right ascension) determination, the basic principle is exactly the same as the "sundial" determination of the sun's RA. Since there are no shadows involved, one observer has to be on a quadrant staircase, and the other on the gnomon staircase, and the objective is for the two of them to align on a star, which will give you its

Fig. 6.2 The author standing in front of the Yantra in Jai Singh's Delhi. The Misra Yantra is visible in the background to the *left*. Though Delhi's observatory still stands, it is in relative disrepair and less well preserved than the observatory of Jaipur. Part of the reason for this is the destruction of much of the observatory during conflict in the rebellion of 1857. In addition to the Delhi observatory, Jai Singh's observatories at Jaipur and Varanasi still stand. The others have been lost

RA position, and define the edge of the shadow that *would* have been cast were the light received from the object bright enough.

The purposes of Jai Singh's observatories were likely numerous. There is almost nothing that has one *single* purpose, and we should take him seriously when he says that he wanted to improve on the existing astronomical tables. But he didn't simply want to do this in any useful or available way. His commitment to reviving ancient Vedic culture, through its rituals and ideals, influenced the way he thought about the sky as well. There was something not quite right about tools such as the telescope. He certainly used it, and noticed what it could do, but he found that it did not suit his purposes. Part of the reason for this was that what Jai Singh, and what classical Indian astronomers before him, wanted to know about the stars, was not necessarily the same as what Western astronomers of his time wanted to know about them. The cosmos is vast, and there are many things to learn about it. Even within our own neighborhood of the earth, there are many different kinds of learning, dictated in part by very different questions. Whether one is engaged in biology, psychology, literature, or sports science has largely to do with what kinds of question one is asking about the human being, for example—with what one is interested in knowing. And the tools of one area will differ from those of another. When we talk about astronomy today, we tend to collapse these distinctions, this

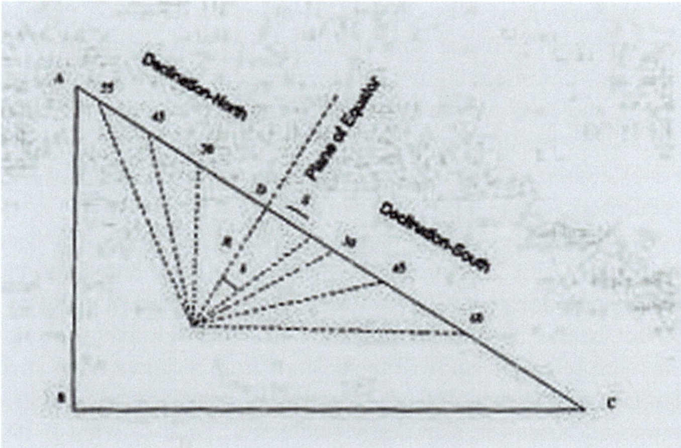

Fig. 6.3 This illustration shows how to determine the declination of an object using the staircase of the of the Yantra. Alignment between an observer on the gnomon staircase and one on one of the quadrants will require two positions, one on each staircase. The position here on the gnomon staircase will determine declination, according to the angles listed here (from Virendra Nath Sharma, Sawai Jai Singh and His Astronomy)

variety of questions and goals, into one, and that one is defined by the particular techniques of scientific astronomy as done in academia and government agencies like NASA. Because of this, other projects involving understanding the cosmos and its connection to or relevance for humanity are rejected as legitimately "astronomical", and are pushed into some kind of strange middle ground. Projects like that of Jai Singh, the Vedic astronomers, the early Maya, or Chinese court observers were not art or literature (though perhaps they had artistic value), but we also reject them as astronomy. Thus sometimes they are deemed "astrology", and left at that. But our conception of astrology is poorly formulated. If it is taken in a weak sense, then it includes everything we deem "astronomy" as well. But if in a strong sense, then it doesn't include most of the systems we've talked about, and becomes a straw man. 'Astrology' is a term most of the traditions discussed in this book did not have, and our projection of it is generally a dismissal, as we view astrology as less serious, less important, and less worthy of study than astronomy.

But let's think about the different senses of this term. In a weak sense, astrology might be understood as a system or systems on which some connection is posited between celestial events and humanity. The human realm has continuity with and is affected by events outside of the earth. This is a commitment of anyone who accepts scientific naturalism. The earth was created out of the accretion disk that swirled around the sun in its early days. Human life is drastically affected by the changes in the sun and the orbit of the moon, which causes tides and precession. The chemical makeup of our bodies, and even the fact that we are alive, is due to facts about the sun and its place, which are in turn due to facts about the formation of the galaxy, and so on. Comets and supernovae, when they appear in the sky, affect us in a number of ways. They can cause panic, stir curiosity, bring us to reflect on the nature of the cosmos. Observing our universe with advanced technology and discovering interesting new things—this affects humanity as well. Astronomy is a human science, and as such what it studies is inevitably connected to humans and affects humans, in multiple ways. So we see that the above weak definition of 'astrology' is not going to work, as it does not differentiate between astronomy and that which we want to call astrology, which takes away the whole point of the distinction.

How about a stronger sense? Perhaps we should call astrological systems those which posit a robust and *unjustifiable* connection between humanity and the cosmos, based on the notion that celestial events have some direct effect over fortune and fate, wealth, health, and the like. This definition would give us a sufficient way to divide the more scientific study of the cosmos from a superstitious investigation. The problem here is that much of what we deem " astrology" due to its difference from contemporary scientific astronomy (including much of what is discussed in this book) does not fit the category by this strong description.

The problem may be this—we want to dismiss or reject the ways of thinking about astronomy discussed in this book, because they are intrinsically different from (though they share some features with) contemporary scientific astronomy, but in order to do this we have to show that they are somehow *irrational* or non-scientific. Merely positing that a pursuit sees a connection and/or continuity between the cosmos and humanity is certainly not irrational, anti-materialist, or non-scientific (any empirical scientist must concede as much). And on the other extreme, attempting to tie ancient and pre-modern conceptions of astronomy to superstitious belief in the direct causation of human events and behaviors by celestial objects and events, which would successfully cordon off these other systems of thought, is ultimately unsuccessful because most of these systems of thought were not "astrological" in this sense.

Indeed, even in our own society we see thought that may be deemed astrological in both senses of the term, and this thought may not be *independent* of the unity-seeking thought that played such a large role in the creation of modern science (as discussed in the next chapter concerning the thought of Johannes Kepler). In addition, commitment to astrology today may be due to the severing of the link between ritual, human spirituality, and astronomy that existed in many of the astronomical systems discussed in this book. Martin Bauer and John Durant argue, in their psychological study of astrological beliefs in contemporary "industrialized world", that:

> …serious interest or involvement in astrology is not primarily the result of a lack of scientific knowledge or understanding; rather, it is a compensatory activity with considerable attractions to segments of the population whose social world is labile or transitional; belief in astrology may be an indicator of the disintegration of community and its concomitant uncertainties and anxieties. Paridoxical as it may appear, astrology may be part and parcel of late modernity.[25]

[25]Bauer and Durant (1997: 55).

So where does Jai Singh fit in this conceptual space, and what does his thought tell us about earlier Indian conceptions of astronomy? Since one of Jai Singh's main goals was to revive Vedic and earlier Indian traditions, and we see this in his focus on ritual as well as astronomy, we can learn a great deal about at least how Jai Singh himself understood the essence of the Indian astronomical tradition. Observation of the sun, moon, planets, and stars with the instruments Jai Singh built itself had a ritualistic aspect. The existence of redundant instruments suggests this. In the Jaipur observatory, there is a small brass Samrat Yantra just beside the large Samrat, on which one can make all of the same observations, and which is about as accurate as the larger one. What is different is the *process* of using these tools, the actions one must take to make the measurements. In much of Western astronomy, and certainly today, the act of observing or measuring or determining coordinates itself is not taken as especially significant—rather, it's what we *get* from our interactions with astronomical tools that is of sole interest, and the observation itself is a mere means. When we consider the meaning of *ritual*, we see that the ritual act itself is accorded significance, independently of any result of the ritual. Certainly this is how many in the Vedic tradition understood ritual. The scholars of the Mimamsa school of Vedic interpretation and others and others stressed the importance of ritual itself for ordering the cosmos, not just for what it could give to humans. Ritual did have as one of its purposes the production of gains for humans, but its most important role was to give structure to and complete the order of the cosmos. This is the basis of the distinction early Indian ritual thinkers drew between *purusartha* (action done for the sake of the human) and *kratvartha* (action done for the sake of ritual). Ritual not only represented, but also brought to completion the order of the cosmos, bringing humans in line with the natural activity of the rest of the cosmos.

The operation of the instruments in Jai Singh's observatories required a kind of ritual human movement that imitated or mirrored that of the celestial objects. In order to find coordinates of a star or a comet on the celestial sphere using the Great Samrat Yantra, two people had to interact together, carry the light of fire, rise up and down a great staircase, and position themselves on a line with the celestial object in question. This process would have had much richer ritual value than the simple process of using the small brass Samrat. The other instruments in Jai Singh's observatories were all built similarly. Since we know that Jai Singh had a great concern with Vedic ritual and revived a number of rituals, we might see the observatories and their use as demonstrating the ritual nature of astronomy in India, which would certainly hearken back to the ancient period. Jai Singh's main goal seems to have been to revive Vedic astronomy and the uniquely Indian way of understanding the sky (Fig. 6.4).[26]

---

[26]Architectural historian Andreas Volwahsen argues in Volwahsen (2001) that the main purpose of the observatories was to mirror the cosmos in architecture. The observatories were akin to *mandalas*. While I think there is something to this—the layout and architecture were certainly *part* of the overall effect, I understand this as an aspect of the larger ritual purpose of the sites, similar to the 360 stone falcon altar of earlier years, or the architecture of a Christian cathedral. The

Fig. 6.4 The grounds of Jai Singh's at Jaipur, which is presently in the best condition of any of the remaining observatories. Numerous instruments can be seen in this photo, including a small Samrat Yantra and Jai Prakash Yantra

References

Anthony, David, 2010, *The Horse, the Wheel, and Language: How Bronze-Age Riders from the Eurasian Steppes Shaped the Modern World*, (Princeton: Princeton University Press)

Baldi, Philip, 1983, *An Introduction to the Indo-European Languages* (Carbondale: Southern Illinois University Press)

Bauer, Martin, and John Durant, 1997, "Belief in Astrology: A Socio-Psychological Analysis", *Culture and Cosmos: A Journal of the History of Astrology and Cultural Astronomy* 1 (1), pp. 55–72.

Griffith, Ralph, 1896, *The Hymns of the Rg Veda*, (Benares: E.J. Lazarus and Co.)

Gupta, R.K. and S.R. Bakshi, 2008, *Studies in Indian History: Rajasthan Through the Ages. Volume 4: Jaipur Rulers and Administration* (Delhi: Sarup and Sons)

Hooja, Rima, 2006, *A History of Rajasthan* (Delhi: Rupa and Co.)

Jamison, Stephanie and Joel Brereton, trans, 2014, *The Rigveda: The Earliest Religious Poetry of India: Volume I* (Oxford: Oxford University Press)

Kak, Subhash, 1995, "The Astronomy of the Age of Geometric Altars," *Quarterly Journal of the Royal Astronomical Society*, 36 (4), pp. 385-395

Kochhar, Rajesh and Jayant Narlikar, 1995, *Astronomy in India: A Perspective* (Delhi: Indian National Science Academy)

Law, S.P., 1986, "The Regulation of Menstrual Cycle and Its Relationship to the Moon", *Acta Obstet Gynecol Scand*, 65 (1): 45-48

(Footnote 26 continued)

architecture is important, but has its significance within the larger context of ritual—the ritual performed at the altar in the falcon case, or the liturgy in the Christian case.

Michaels, Axel, 2004, *Hinduism: Past and Present*, (Princeton: Princeton University Press)

Mignolo, Walter, 1995, *The Darker Side of the Renaissance: Literacy, Territoriality, and Colonization* (Ann Arbor: University of Michigan Press)

Milbrath, Susan, 1999, *Star Gods of the Maya: Astronomy in Art, Folklore, and Calendars* (Austin: University of Texas Press)

Nasr, Seyyed Hossein, 1987, *Science and Civilization in Islam, Second Edition* (Cambridge: Islamic Texts Society)

North, John, 2008, *Cosmos: An Illustrated History of Astronomy and Cosmology* (Chicago: University of Chicago Press)

Rao, S. Balachandra, 2000, *Indian Astronomy: An Introduction* (Hyderabad: Universities Press)

Rao, S. Balachandra, 2010, "Classical Astronomy in India—An Overview", in Padmanabhan, (ed) *Astronomy in India: A Historical Perspective* (Delhi: Springer [India]).

Sharma, Virendra Nath, 1995, *Sawai Jai Singh and His Astronomy* (Delhi: Motilal Banarsidass)

Subbarayappa, B.V. and K.V. Sarma, 1985, *Indian Astronomy: A Source-book* (Mumbai: Nehru Centre)

Thapar, Romila, 1997, *Early India: From the Origins to AD 1300* (Berkeley: University of California Press)

Volwahsen, Andreas, 2001, *Cosmic Architecture in India: The Astronomical Monuments of Jai Singh II* (Munich: Prestel)

Chapter 7
Eclipse—Shadows and Light

We have seen the significance of solar and lunar eclipses in India and other cultures, and looked at the attempts to calculate these events and observe them and other celestial events with great precision by building enormous tools. But how do we understand these breathtaking phenomena in our contemporary scientific understanding?

Solar and lunar eclipses are two versions of the same phenomena—flip sides of a single coin, if you will. Which, consequently, is also the same phenomenon as the transit of a planet across the sun, or of any astronomical object across any other. The obscuration of one astronomical object by another, referred to as a *transit*, is at the heart of the explanation of eclipses. A transit happens whenever one object moves in front of (passes across) another in our line of sight from the earth. Of course, from some vantage point any object in the universe can be seen passing in front of any other, so the notion of a transit is dependent on one's location and line of sight. All of us on earth occupy basically the same point in space relative to the distance of astronomical objects, so we will all see transits in the same way. Of course, this changes depending on the distance of the objects we view and the place on earth we observe from. For deep-space objects, everyone on earth will view transits in the same way. For transits that happen closer to our own neighborhood, the situation is a bit more complicated. Solar eclipses are the *most* local kinds of transit widely observed (only near-earth asteroid transits might be closer), because of the proximity of the moon to the earth.

A solar eclipse is of course a transit of the moon across the sun. Because of the nearness of the moon, people at different places on the earth will see the transit differently, and some will not see a transit at all. Solar eclipses are thus one of the only kinds of transit (and the only widely observable and without astronomical tools) not visible everywhere on earth the transited object (in this case the sun) is visible. Because the moon is so close to us, it appears at very different points in the sky depending on where on earth one observes it from. One will only see a transit at all if one is within the zone on earth in which the angle of the view from the observer, moon, and sun is enough to put the moon somewhere over the disk of the

A. McLeod, *Astronomy in the Ancient World*,
Historical & Cultural Astronomy, DOI 10.1007/978-3-319-23600-1_7

sun from the observer's vantage point. For any given solar eclipse then, this will only happen in a relatively narrow area within which the observing angle is within the arc of the sun. This is why solar eclipses, when they happen, are not visible everywhere on earth. When in New York City observers see a brilliant total eclipse of the sun, in Buenos Aires at the very same time observers will see the full arc of the sun shining brightly as on a normal day. For this same reason, even those who are able to see a solar eclipse from their vantage point on earth will see it differently, depending on where along the path of obscuration they are (Fig. 7.1).

There are different types of solar eclipses, which I will get into further below, but for the moment I will consider *total* solar eclipses. A total solar eclipse is one with a path such that at some place on earth the moon will transit the sun perfectly—that is, from some vantage point on earth the center of the moon will pass directly over the center of the sun. There are different ways this can happen, different paths of totality that can be taken (which accounts for the different locations from which a total solar eclipse is visible). At the center of a path of totality, an observer will witness a total solar eclipse, one of the most magnificent astronomical events observable from earth. Because of the strange coincidence that the disc of the sun

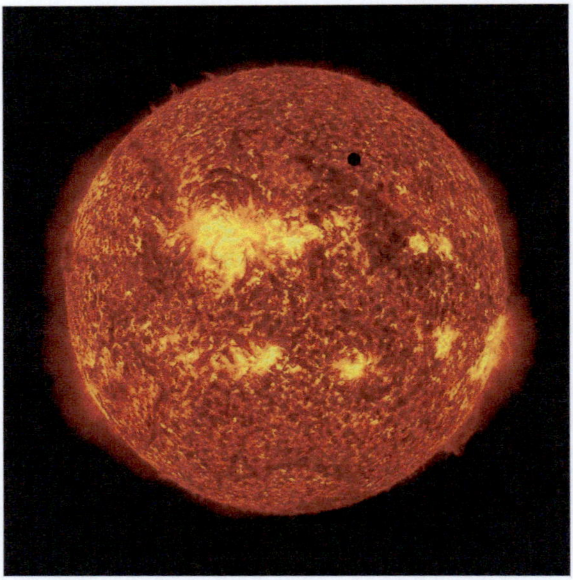

Fig. 7.1 Transit of Venus across the sun, 2012. The orbit of the planet takes it over the surface of the sun from our standpoint on earth. Both solar and lunar eclipses are created by the same phenomenon. In a solar eclipse, the moon transits the sun. In a lunar eclipse, it is the earth that transits the sun, from the perspective of the moon. The darkening of the moon we see in lunar eclipse the shadow of the earth falling over the moon from this transit. From the lunar surface, what we see as a lunar eclipse would resemble a total solar eclipse (though the disk of the moon is much larger than that of the sun from the perspective of the moon) (Photo credit: NASA (Solar Dynamics Observatory))

appears almost exactly the same size as the disc of the moon from earth because of its distance and the much smaller moon's close proximity, the moon's apparent disk will seem to fit exactly overtop that of the sun, covering it completely but not extending beyond it (this is not always the case with eclipses though, as we will see below). The overpowering brightness of the sun withers to a faint glow from its coronal edge shining outward from behind the blocking disk of the moon.

Not all solar eclipses are total, but during a total eclipse not all vantage points from which the eclipse can be seen will reach totality. Totality is defined along a path within which observers will see the full covering of the sun. Outside of this path, observers will either see a partial eclipse or no eclipse at all, depending on where they observe from. Those closer to the path of totality will see a greater portion of the sun eclipsed, and the further one gets away from this path, the less of the sun one will see eclipsed, until outside of the shadow of the moon completely one will see the sun as bright as on a normal day (or not at all, of course, if one is on the night-side of the earth at that time). While in the path of totality one can glimpse upon a total eclipse and see the faint coronal outline of the sun, hundreds of miles to the south one may only witness the moon arcing across half of the sun, a glancing blow, what we call a *partial* eclipse. The full disk of the moon will never pass directly in front of the sun as in totality, but will instead pass over only a part. Thus, what for some observers along the path of totality will become a total solar eclipse, for other observers will only be at most a partial eclipse (Fig. 7.2).

Not all solar eclipses are total somewhere on the earth's surface. There are some eclipses that have no path of totality and appear partial wherever on the earth they are seen. These are properly called *partial* eclipses, in distinction from total eclipses that are observed as partial outside of the path of totality. In partial eclipses, there *is*

Fig. 7.2 Sequence of an annular solar eclipse, May 20, 2012 (Photo credit: By Brocken Inaglory (Own work))

no path of totality. How, one might ask, could such an eclipse happen? If we think back to the idea of sight-lines we can make some sense of this. For any two bodies, whether in space, in the air, or wherever, there is some point outside of both that intersects both when a straight line is drawn between the three (there will actually be more than one such point but we'll put that aside for purposes of this—a point on the opposite side of the two objects also intersecting the same line will give a flipped transit, and also every point along the line intersecting the two objects will be a totality point. Since we are considering the "plane" of the earth roughly perpendicular to the line in question, though, there is no need for us to consider points beyond this plane). Now, since the "plane" of the earth is moving (as the earth rotates), think of the point along the plane where totality is visible as extended across the plane moving opposite the direction of motion of the plane. The places that will achieve totality are those along this new line (the actual path of totality will itself be an extended plane, but we can leave that aside here). As mentioned above, a region above and below this path of totality will receive only a partial eclipse, until one gets too far from the path and no eclipse is visible at all. A partial eclipse happens on earth when the path of totality here (for the theoretical observer in space) passes just above or below the earth, enough for the shadow of partiality to pass over the planet. In such a case, were we to travel out into space just below or above one of the poles, we would see a total solar eclipse. Of course, this is not very special, since we could at any time travel to some point in space close to the earth and see a total solar eclipse, simply by putting ourselves in some point on the far side of the moon from the sun along the line intersecting the sun and the moon. We would never have to get very far from earth to do this. For the path of totality to cut across the earth is less frequent (although still relatively frequent). To have totality in any *given* spot on earth, however, is fairly infrequent. The United States has not experienced a total eclipse since that of 1991, visible from Hawaii. There have been none visible from the continental US since 1979. The next to appear in the US will be in 2017.

Another kind of eclipse is the *annular* eclipse, which might be seen as the younger brother of the total eclipse. In an annular eclipse, the center of the moon's disk does move through that of the sun's disk, but because the moon is further in its elliptical orbit around the earth, the apparent size of the disk is smaller than that of the sun, and the moon's disk does not completely cover the sun's disk. At the peak of such an eclipse, them, a bright ring of sun is still visible surrounding the dark disk of the new moon. There is a monthly change of about 4.07' of arc between the maximum and the minimum size of the moon depending on the distance from earth in its orbit (the maximum size coming at closest approach and the minimum at furthest). Thus given that for a total eclipse the moon needs to be relatively close to us in its orbit and needs to be aligned so as for the path of totality to pass over the earth, they are not as common as one might expect.

There is also an additional complication to explain why total eclipses (or eclipses in general) do not happen much more often. If the moon orbited the earth around its center, then we would experience an eclipse every time the sun crossed the plane of the celestial equator where the moon sat—which would give us far more eclipses. If we went one step further and straightened out the rotation of the earth so it were no longer tilted on its axis and the celestial equator then lined up perfectly with the ecliptic, we would experience a total eclipse about once a month!

While some people in ancient India thought that the obscuration of the sun by Rahu and Ketu (either a disembodied head and its body or two invisible planets) was responsible for solar eclipses, other more astronomically sophisticated people understood that the lunar nodes played a role in the regularities of solar eclipses—a regularity that is difficult, but not impossible, to discover. The main problem is the amount of time and attention it takes to recognize the patterns of eclipses—one needs multiple decades of observation to begin to detect the patterns of solar eclipse. Astronomers in India knew about the patterns of eclipse from relatively early on. Some referred to the lunar nodes as Rahu and Ketu. Lunar nodes are one part of the explanation of why eclipses happen as they do, and their pattern and frequency.

The earth is tilted 23.5° on its axis, so that the celestial equator is offset from the ecliptic (roughly the plane of the planets and the sun) by this many degrees. So the moon does not orbit around the equator of the earth, as envisioned in the scenario above, but will appear either high or low in the sky depending on its position relative to an observer. In addition, the orbital plane of the moon is not exactly aligned to the ecliptic (roughly the plane of the planets' orbits around the sun), but is elevated 5° to the ecliptic. This means that the moon can appear anywhere from 28.5° above the celestial equator to 28.5° below the celestial equator in the sky (5° on either side of the ecliptic). Since the orbit of the moon is elevated to that of the earth around the sun (the ecliptic), there will be exactly two points along the orbit of the moon at which it intersects the ecliptic. These are the lunar nodes. A solar eclipse can only happen when the moon is at a lunar node. The reason for this is clear enough—the sun only "moves" along the ecliptic, since this plane is defined by the earth's orbit around it, and the moon can only transit the sun, or cross over its face, when it also occupies a point on the ecliptic. Notice though that the moon will cross over a lunar node two times every month. There are not, however, two solar eclipses somewhere on earth every month. Why is this? (Fig. 7.3).

In order for there to be a solar eclipse, not only does the moon have to be at a lunar node, but it also has to be at or very near new moon phase. This is because it is only during this phase that the moon is facing the sun. If the moon achieves a lunar node point at some other phase, it is not facing the sun from the vantage point of the earth, and so there will be no transit. Although the lunar nodes are at the same places relative to the position of the earth, since the earth moves around the sun those positions will not always correspond with the same phase of the moon. On January 1st, for example, say that the moon is at a lunar node and at full moon (on the opposite side of the earth from the sun side). A month later, then, as the earth has traversed 1/12th of its orbit, the moon reaching the same lunar node will now appear in 1/12th less than full gibbous. After three more months, the moon will be in quarter phase during the same node crossing. Three months later, the moon will cross the node in new moon phase. This creates an eclipse season. Such a season, in which eclipses are possible, will last about 34.5 days, and will happen twice a year,

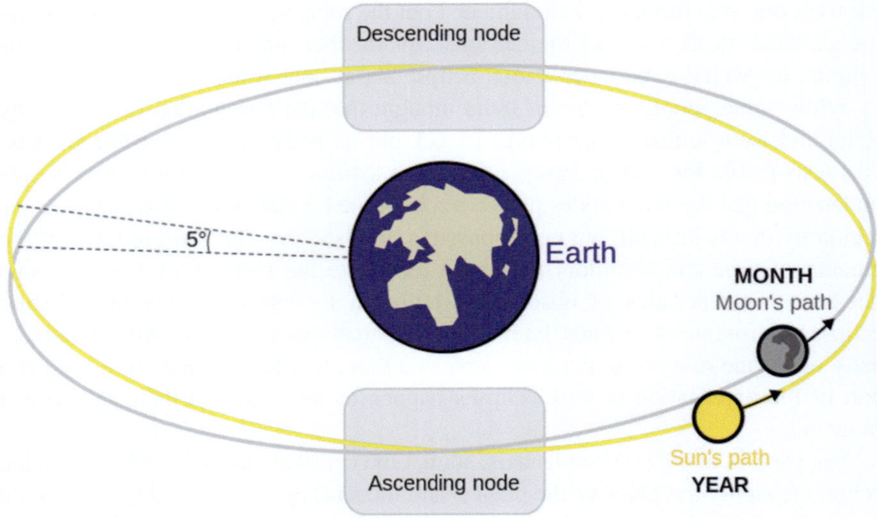

Fig. 7.3 One understanding of "Rahu" and "Ketu" in early Indian thought was as representing the lunar nodes, which are the only points at which eclipses are possible. The nodes are formed by features of the orbit of the moon around the earth and the rotation of the earth. As can be seen in this illustration, the orbit of the moon is elevated to the plane of the earth's rotation (as well as the ecliptic). There will be two nodes on any given pass of the moon around the earth—one on its way above the plane of rotation of the earth (ascending node) and one on its way below this plane (descending node). Because these two points are the only places the moon can pass through the plane through which the sun appears to us to pass, they are the only points eclipses are possible

once for each node—the ascending and descending node. The ascending node is that through which the moon passes on its rise above the ecliptic, while the descending is passed through on its move below it. Both solar and lunar eclipses happen during these seasons. Because the lunar orbit and the earth's orbit around the sun take different amounts of time and do not form a perfect fraction (365.24 days of the tropical year divided by 27.3 days, the sidereal period of the moon, is 13.38, thus the moon completes 13 orbits and one partial orbit each year), the eclipse seasons will shift from year to year. Thus, if there are eclipses early in January of this year, they will move to a later date in the following year.

There are some regularities that can be discovered by the careful and enduring observer. There are a number of different eclipse *cycles*—collections of periods that repeat in predictable patterns, and during which eclipses (both solar and lunar) will happen. The best known of these cycles, and one which was known to many early civilizations in Asia and Mesoamerica, is the *saros* cycle. The saros is a period of time calculated between eclipses at about 18 years (and 11 days) that can be used to predict eclipses because this is the period of time from one alignment of earth, sun, and moon, for them to occur in the same alignment again. The saros can be calculated independently of eclipses—the configuration of the earth, sun, and moon against the background stars that exists on the day you read these words, for example, will reoccur roughly 18 years and 11 days from now. In the case of most days, this cycle is not very interesting. But for events like eclipses, it can be very useful. If I have been able to calculate the saros cycle accurately, then I should be able to predict when the next eclipse will take place in the location at which one is observed. If a solar eclipse happens today with a path of totality running overtop of my house in northern Colorado, I know that in 18 years and 11 days, another such eclipse will take place, with a path roughly the same over the sphere of the earth. The next eclipse will very likely not pass over my house in exactly the same way as the first, if at all. It is as likely as not that I may not see the next eclipse in the saros cycle *at all*! Why is that? Well, although the 18 year and 11 day saros cycle determines when the earth, moon, and sun will be similarly aligned again, this does not take into account the rotation of the earth. Because the saros cycle is not *exactly* 18 years and 11 days (it is actually 18 years, 11 and 1/3 days), the next time an eclipse in this saros series appears, the earth will be advanced in its rotation 1/3 of a day beyond where it was when the current eclipse appeared. Thus, the path of totality of this new eclipse will be 1/3rd of the way around the earth west of the path of the initial eclipse.

We can see an example of this in two of the total solar eclipses that are part of Saros series 145—the eclipse of August 11, 1999, and the forthcoming eclipse of August 21, 2017. In 1999, the path of totality began in the Atlantic Ocean off the coast of eastern Canada as the sun rose, and stretched across northern Europe and south into the middle east and India before ending just off the coast of India at sunset. The center of the path of totality, the point at which there will be the greatest eclipse, was in eastern Europe. In the August 2017 eclipse, this point will be in the central United States (around southern Illinois), with the path beginning in the mid-pacific and ending deep in the Atlantic almost to Africa. The greatest eclipse point in 1999, in Romania, happened around 29° E, while the greatest eclipse point in the 2017 eclipse (and thus the longest eclipse time) will happen at around 89° W longitude. This yields a 110° difference between the two points (89 + 29), which, divided by 360°, gives us just about 3—the 1/3rd of a day added to the 18 years and 11 days of the saros cycle (Fig. 7.4).

The saros is not the only cycle that can be useful in prediction of eclipses, even though every eclipse can be calculated as a part of some saros cycle. There are a few other interesting features of saros cycles. There is, of course, far more than one saros cycle, as the alignments necessary for eclipses to occur happen much more frequently than the return of the earth moon and sun to any *particular* alignment. Remember that there are two eclipse seasons in every year, corresponding to the moon's passing through the two lunar nodes at points when alignment with the sun and earth is sufficient to create eclipse. Given that the length of a saros period is roughly 18 years, then beginning with one eclipse, every eclipse that happens between that eclipse and the next one in its saros cycle 18 years later will be part of its own saros cycle. Given that there is an average of 2.4 solar eclipses in a year, we

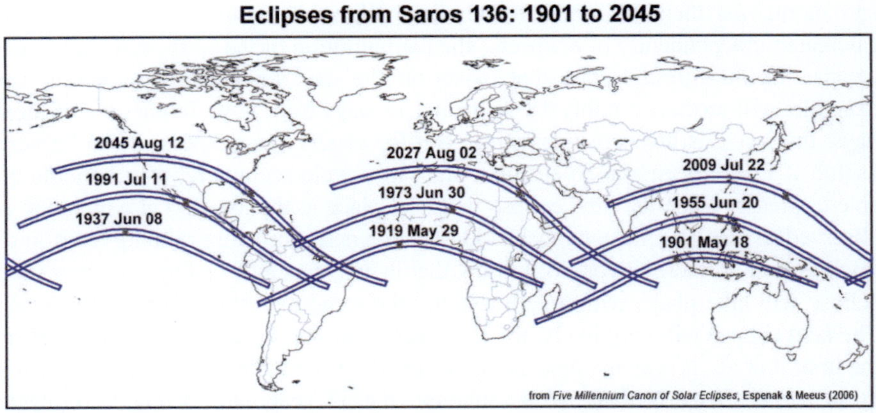

Fig. 7.4 Eclipse paths of the nine total solar eclipses of (solar) Saros 136 between 1901 and 2045. The eclipses of Saros 136 will achieve totality until 2496, then returning to partial eclipses until the end of the series. The next eclipse in this series will take place in August of 2027. Observers in the continental United States will be treated to the first total solar eclipse visible from the region in 47 years with the total solar eclipse of August 2017. This eclipse is part of Saros 145 (Photo credit: NASA (Espinak and Meeus, "Five Millenium Canon of Solar Eclipses" −1999 to +3000))

multiply this number by 18 (years), and this yields the number 43.2 eclipses. And this is just what we find for the saros cycle—there are about 40 active saros cycles of eclipses at any time. But what do we mean by an *active* saros cycle?

Any given eclipse that happens through the year during the eclipse season will have a corresponding saros series. It may be first, last, or somewhere in the middle of the series. For those eclipses at the edges of the cycle, there will be certain features. All saros cycles begin and end as partial eclipses, near the poles, as the sun enters the lunar node. Whether the series begins near the north or south pole depends on whether it is a cycle of an ascending or descending lunar node. Those associated with the ascending node will begin in the south, while those of the descending node will begin in the north. While, as explained above, the path of totality of the eclipses in a given saros cycle (when there *is* a path of totality) will move westward about 1/3 of the earth's circumference every time they appear, the path will also move either northward or southward for each eclipse of the cycle, although not as dramatically as the westward movement due to the earth's rotation. This change in latitude happens slowly over about an average of 71 appearances of eclipses in a cycle (although the number of eclipses in any saros cycle can vary slightly). Saros cycles last a long time on any measure, as if we take the average of 71 eclipses, minus one for the initial eclipse, multiplied by 18 years, we yield 1260 years (though a series with a greater number of eclipses can last up to 1550 years). The eclipse of August 2017, which will be visible in the United States, for example, is a mid-cycle eclipse of a descending node saros cycle (Solar saros cycle 145) that began with an eclipse near the north pole in January 1639, and will end with an eclipse near the south pole in April 3009.

One feature of this saros cycle that is shared by all others is of relevance here. The first 15 as well as the final 19 of the eclipses of this cycle are only partial eclipses. As we saw before, the reason for this is that the path of totality in these eclipses will be off the surface of the earth due to the movement toward the poles. In the first eclipses of the series, the shadow will graze the earth near the pole, and in the final eclipses the same will happen near the opposite pole as the shadow retreats. The reason that the paths move either northward or southward and that there are thus a limited number of eclipses in a saros cycle is that the alignment between the earth, moon, in the same configuration once every 18 years 11 and 1/3 days will be slightly different against the background of stars, and this will cause precession of lunar nodes from the perspective of the earth, thus accounting for the changing paths and the finite existence of each saros cycle.

As mentioned above, there are other cycles that can be useful in tracking and predicting eclipses, including the tritos cycle. It is also the case that *lunar* eclipses follow cycles similar to those of solar eclipses. There are lunar saros cycles, for example, that work in the same way as the solar cycles. This is because the principles behind solar and lunar eclipses are the same, despite some important differences that will make a big difference to observers. Lunar eclipses are far more common for any given observer on the earth, because any time a lunar eclipse happens it will be visible from any place on earth from which the moon is visible. There is no narrow path of totality as is the case in solar eclipses, because since a lunar eclipse is created by a transit of the earth across the sun from the perspective of the moon, any observer on earth who can see the moon will see the shadow of the earth fall over the moon in whatever way it does, either partially or fully covering it.

The observation of eclipses, and transits more generally, has played an enormous role in advancing our understanding not only of the science behind eclipses, which is relatively straightforward and involves nothing more than planetary motions and trigonometry.

Eclipses in History

The ritual nature of astronomy in India not only influenced how people made observations and devised theory, but also how they interpreted celestial events. While in the case of early China celestial events as *portents* is often stressed, in the case of early India there is a somewhat different understanding of rare and spectacular celestial events. The notion of celestial events as portents is not completely absent from early Indian thought, but the focus, in ritual context, is more on acting *at the proper time*. And it is often a celestial event that can signal to us when that proper time is. Ritual in general concerns codified actions. Rituals are *performed* rather than serving as the content of knowledge. We can know *how* to perform rituals, but the full meaning of the ritual itself is expressed in its performance. Various ritual schools in early India, including the *Mimamsa* school of ritual interpretation, reduced even philosophical questions to the practice of ritual itself. All we need to do, according to Mimamsa thinkers, is to interpret the world so as to understand ritual requirements and proper practice.[1] Proper performance of ritual will ultimately lead us to the main human goal, *svarga* (heaven), which Mimamsa thinkers understand not as a transcendent state or something attained after death, but rather happiness or satisfaction within this life.

Solar eclipses, like other celestial events, can be seen as signaling the proper times for certain ritual actions. We see accounts of such understandings in a number of Indian texts. A number of accounts from the 11th century CE list people who gave gifts or land grants on the occasion of solar eclipses. These were seen as auspicious events, rather than harbingers of doom. Both solar and lunar eclipses held this association for many people in premodern India.[2] The concerns of the astronomer were to discover facts about the skies that would help determine the most auspicious time to perform rituals and to order one's space of living. As we have seen in the example of Jai Singh's observatories, what was most important in early Indian astronomy was not data to bolster theoretical understandings (and indeed this was not important in the west either for most of its history). Instead, corresponding to the cosmos in correct ritual patterns formed the basis of Indian astronomy. Ritual is not only based in the world, but ritual *orders* the world. That is, the human, in performance of ritual, takes part in making the world as it is. We are not simply passive entities, but take a constructive role in the cosmos. It is within ritual, then, that humans find their proper place in the world as a whole, not only in their community. The *Mimamsa Sutra* of Jaimini (3rd c. BCE) offers a clear statement of this:

> The result [of ritual]…is for the sake of *purusa* or the individual soul [person]; and the soul too is for the sake of action; and these are connected together by means of purpose. This is a universal law, applicable to all cases and in all circumstances; for if there be no purpose, there can be no action. The result of an action follows action: were it otherwise, we should be able to get the result without action; and a person is moved to action because of the effect or result he expects to produce. The purpose and qualities of a person become one in action, for that is the law of action.[3]

Astronomy in India, perhaps more than in any of the other cultures included in this book, was a *performative* area of thought—whether that performance was mathematical, as in the work of the Siddhantic astronomers, or physical, as in the case of Jai Singh and his revival of ancient Vedic ritual.

[1]"Although Mimamsa does not enter into any philosophical analysis of the universe, it welcomes all philosophical discussion that will further an understanding of right action as enjoined in the Veda." (Bernard 1947, 103).

[2]A number of texts claiming the auspiciousness of eclipses and particular examples of ritual action in connection with eclipses are discussed in (Misra 1973, 72).

[3]*Mimamsa Sutra* 3.1 (Thadani 2007, 43).

References

Bernard, Theos, 1947, *Hindu Philosophy* (Delhi: Motilal Banarsidass)
Levy, David, 2000, *Eclipse: Voyage to Darkness and Light* (New York: ibooks)–2010, *David Levy's Guide to Eclipses, Transits, and Occultations* (Cambridge: Cambridge University Press)
Misra, Vibhuti, 1973, *Religious Beliefs and Practices of North India During the Mediaeval Period* (Leiden: Brill)
Thadani, N.V., 2007, *Mimamsa Sutra of Jaimini* (Delhi: Bharatiya Kala Prakashan)

Part IV
The Perfection of the Spheres

Throughout this book thus far, I have offered a comparison between the astronomical ideas of ancient (and some more modern) cultures outside of the West and contemporary ways of understanding celestial events. Although these contemporary understandings are not solely the domain of the West, much of the conceptual underpinning of these understandings *are* derived from Western thought. Astronomy today has become truly a global phenomenon—people working on the understanding of galaxies, supernovae, black holes, dark matter, and all kinds of other astronomical issues, come from countries around the world. Traditions that were once very separate, such as those of China, India, and Europe, now work on the same problems in the same ways, often working together. The European Space Agency recently sent a groundbreaking mission to study a comet, while the Indian Space Research Organization only months before put an orbiter around Mars. The China National Space Administration has also been involved in space exploration and scientific research, sending missions to the moon. All of the data gained on these various missions is used, interpreted, and understood using the same methods, the same tools, and the same background assumptions. Today, we might say that astronomy has become globalized, homogenized, and professionalized. Part of this, of course, is due to more general globalization, which has connected us all in ways we have never been connected before. But it is also undeniable that much of this globalization has been one way, as "Westernization." Although the ancestry of contemporary astronomy is not limited to the West (a fact not often noted in histories of astronomy), a large part of its background originates in Western thought, and many of the philosophical assumptions underpinning the scientific focus of contemporary astronomy come from the West.

Every comparison is an investigation of different ways of understanding the world, no one of which has title to definitively claim itself as a representative of "the way things are." Thus, when I have compared ancient and contemporary understandings of astronomical events in previous chapters, I have not attempted to do so from a place of comparing ancient thought against the background of "the right view." Rather, I have compared these ancient views with the ways that astronomers in the contemporary world think about these events. Are either of these the *true* understanding? This is impossible to say, given what we know (and do not know) about the universe in which we live. Every way of understanding the world comes with its own presuppositions, assumptions, and concerns. The accounts we build of the universe are not objectively better, or more complete, than those we have looked at from the ancient world. The contemporary story of astronomy is still a story, more or less significant and accurate in that it expresses to us some important aspect or element of the cosmos in which we live.

Thus, in this chapter I look at astronomy in the ancient and early Western world. I discuss some of the views and assumptions that have informed its astronomy and continue to inform our understanding of the cosmos today, even though they are often unrecognized. In the final section, I describe the ways in which the early Western views developed into what we know today as astronomy. Comparing ancient views with those of our modern understanding will reveal hidden assumptions and philosophical commitments.

Chapter 8
Europe and the West

Pre Greco-Roman Astronomy in Europe

While the "Western tradition" is often traced back to the Greeks, and thus in many histories of Western astronomy the first word is given to the ancient Greek philosophers and astronomers, there is a much older tradition of concern with the sky in Europe, dating back to the earliest periods of human culture in the region.

Stonehenge

One of the most famous astronomical sites in the world sits in the present day southern part of England, in the county of Wiltshire, on a vast open plain. The megalithic period rock structure today known as Stonehenge has fascinated visitors and readers for many years, and it has existed in its present location, in something similar to its present state, for over 4000 years—although it was much more complete of a structure when it was built around 2000 BCE. The present day Stonehenge represents the "ruins" of the site, in much the way that the Roman Coliseum as it stands today gives us something of the sense of the original, but is largely the skeletal bones of the original structure (Fig. 8.1).

It is believed that the present rock structure identified with Stonehenge is actually a monument to an even earlier site with astronomical significance, and which used the sight lines of the later stone site.[1] Likely the earlier site developed over many years as did Stonehenge itself, which evolved and changed for over 1000 years before reaching its final state, the remnants of which stand today. Also, as is not well known, Stonehenge is only a part of what was a larger site on

[1]This position was argued for by Owen Gingerich (1975), and recent discoveries seem to lend further support to the position.

A. McLeod, *Astronomy in the Ancient World*,
Historical & Cultural Astronomy, DOI 10.1007/978-3-319-23600-1_8

(a)

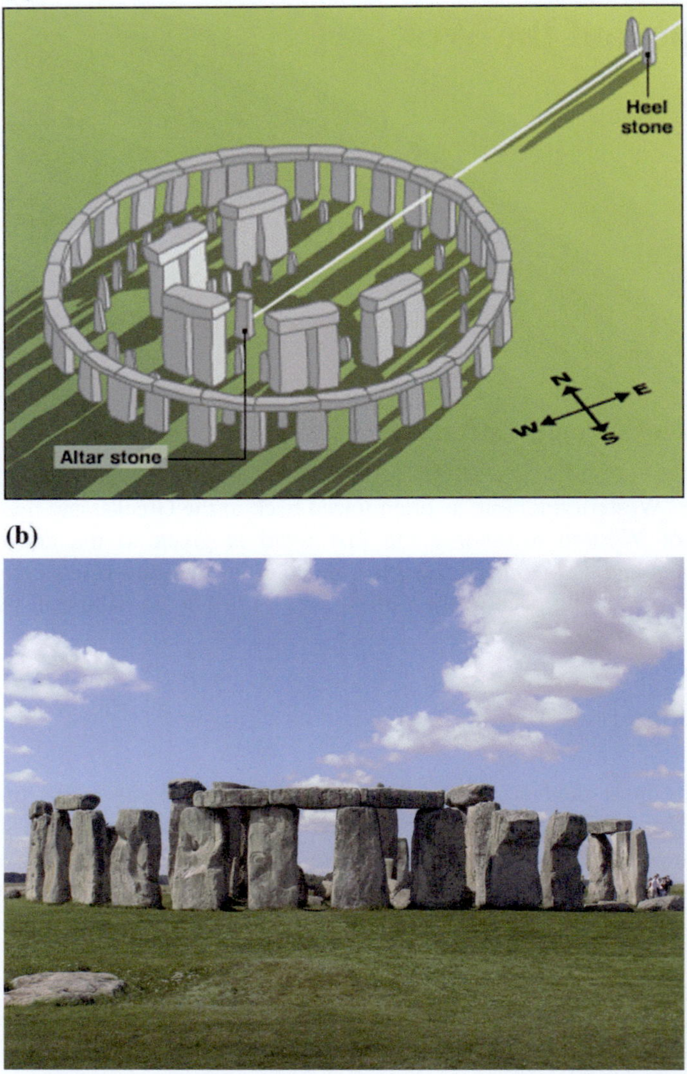

(b)

Fig. 8.1 Drawing of how Stonehenge likely looked during the early years after its construction (*left*), and a photo of Stonehenge today (*right*) (Photo credit: Gareth Wiscombe, Creative Commons https://www.flickr.com/photos/garethwiscombe/1071477228/in/photostream/)

Stonehenge's region of what is now known as the Salisbury Plain. A number of other nearby sites, including Woodhenge, were constructed to align with certain important celestial objects and events, such as the heliacal rising of important stars and planets. Recently uncovered monuments nearby (mere months before the writing of this book), discovered underground at both the Stonehenge site and the

Durrington Walls site, include a number of pillars and a burial house predating Stonehenge by some 3000 years. These discoveries show that the Stonehenge site was much more extensive than previously believed, and that there indeed may have been astronomical involvement with the site long before Stonehenge was built, as there was certainly religious significance, given the burial house.

The visible construction of Stonehenge itself is comprised of a number of different parts and layers. The rock we see, that looks relatively uniform, is actually made up of different material in the different layers. The outermost circle (only about a third of which is still standing) was made of sarsen sandstone, forming a circle with a 100 foot diameter. Inside this circle is a second, or dolerite bluestone, forming a circle (again, not complete in its present state of disrepair) with a 75 foot diameter. Within this are a number of different components, including two "horseshoe" shaped structures and a central altar. The famous "pi" shaped rock structures forming the outer and inner circles were originally connected by overhanging rocks that connected each to one another, some of which still exist intact. The construction of these structures involved creating grooves in the bottom of rock fixtures, designed to fit protrusions created on the mounting rocks, so that the top rocks could basically be "plugged" into the mounting rocks rising from the ground. Each of the top layer rocks were also connected to one another by a similar process, though one that did not involve "plugging", but rather "ridging", creating an indentation in the end of a rock and a protrusion on the other that allowed one to be slid into the one beside it. The reason this was necessary was that, unlike the connection to the mounting rocks, where the main stress would be outward along the plane parallel to the ground where the two were connected, the main stress in the connection between the top rocks would be downward along the perpendicular plane connecting them.[2]

There is still much that is unknown about the Stonehenge site (as is unsurprising for a construction of such deep antiquity with no connected written record and very few material artifacts. The site likely had numerous purposes, rather than one single purpose, as was the norm for ancient sites, and is still the norm today (other than in specific specialized sites and instruments—a phenomenon that is relatively new in human history). One of these purposes was likely astronomical.

Just as we saw in the case of "Cahokia's Woodhenge" and the Fort Ancient town today called SunWatch, Stonehenge seems to have been constructed so as to take advantage of certain important alignments along the horizon, including the summer and winter solstices, and the heliacal rising of certain stars and clusters. It is unclear if astronomical alignments were connected to the main (or an important) purpose of Stonehenge, or if it was instead a minor aspect of the site. Or it could have been unintended to have *any* astronomical significance whatsoever! It is possible that the alignments various scholars claim to have found at Stonehenge could be completely coincidental. Any circular monument or construction on the model of Stonehenge would have alignments with some important celestial event or other, but this does

[2]Chippindale (2004, 10-19).

not mean that such alignments were intended. Take as a case in point the layout of the city of Chicago, whose streets are laid out according to the cardinal directions. Streets run west to east, and north to south. In downtown Chicago, there are a number of enormous skyscrapers that block out great swaths of the horizon and much of the sky overhead as well. But if one stands on one of the east-west streets at either of the equinoxes, facing east at sunrise or west at sunset, one will see the sunrise or sunset exactly between the large buildings—that is, due east or due west. But this alignment is just due to the fact that the sun rises exactly east and sets exactly west on the equinox, and the roads in Chicago happen to be set up east to west. This coincidence makes this alignment possible, but the planners of the city of Chicago did not intend to mark this alignment when they decided to plan the streets following the cardinal directions. More likely they thought that this would make navigation of the city much more convenient than if they laid it out some other way. Some have even coined a joke term to describe the alignment in Chicago at the equinox: "Chicagohenge". Imagine future archaeologists thousands of years distant discovering the ruins of Chicago—would they take this alignment to signal some important astronomical purpose of the city of Chicago?[3]

In addition, if Stonehenge was did have an astronomical purpose, as at least one of its purposes, it need not have been a major purpose. Astronomical interest can be (as many contemporary astronomers can attest) superseded by other concerns that people find more pressing, urgent, or interesting. The stain-glass windows of a cathedral do indeed have the purpose of reflecting light in particular ways that draw the eye, and illustrate artistic scenes, but this purpose is not the main purpose of a cathedral, but merely one among many, and a relatively minor one at that, given what the central functions of a cathedral are meant to be. Given recent unearthed evidence, it seems that Gingerich's view that Stonehenge was mainly a monument to earlier observatories may be plausible. It may indeed be the case that although Stonehenge mimics an observatory, that it was never intended to be used as such. Why might we think this? First, if the ancients knew anything about precession (which is highly controversial but certainly possible), they would have recognized that circle observatories could only be used for so long, and would eventually become obsolete. More realistically, the fact that most of the observatories made around this time and at this site were wooden and inherently disposable or of limited durability makes Stonehenge unusual and unique. Building a structure with stone like Stonehenge clearly shows that the builders intended the site to be permanent in a way the wooden observatories could not be, so they intended for it to exist indefinitely, or at least far into the future—to a time they may have known it would not be useful as observatory (as their other observatories were more perishable). Second, recently uncovered evidence shows that there were burials nearby, and that the site

[3]There is a similar phenomenon in New York City people refer to as Manhattanhenge, but this seems to me more contrived, as the sunrise and sunset alignments in Manhattan don't closely correspond to any particularly significant solar placement, such as the solstices or equinoxes. This is the case because the streets of Manhattan are aligned with the angle of the island, rather than cardinal directions or solstice sunrise/sunset directions.

likely began as a burial ground.[4] This suggests that Stonehenge may have been a monument to the dead. Perhaps Stonehenge was a burial place for priests, astronomers, or rulers, and the Stonehenge site is something akin to an ancient version of the massive gravestones one sometimes finds in the most prominent places in graveyards, or the enormous monuments to rulers or important religious figures.

But Stonehenge largely remains, like much very ancient material, a kind of Rorschach test, saying as much, or perhaps even more, about us and our own interests than it does about the ancients. People have seen in Stonehenge things as diverse as an astronomical observatory, a religious ritual center, a Roman amphitheater, and many other things. The mysteries still surrounding the site may never be solved, but if there is astronomical significance of some extent to the site, it shows that peoples in the early years of European history, well before Roman influence, were concerned with horizon astronomy as we see in many other cultures. The reasons for such interest, of course (and as we have seen) are fairly obvious—connection to growing seasons and maintenance of the calendar. Yet, as we have seen with cultures like that of the ancient Maya, there may have been additional reasons for such a concern, including religious and political ones.

There are other early European sites of even greater antiquity than Stonehenge, throughout the continent and even in the islands of Great Britain and Ireland. One of the oldest of these, thought to have much the same astronomical purpose as sites like Stonehenge or SunWatch, is the large earthen circle from the Neolithic period in Goseck, Germany, known as the "Goseck Circle".

Goseck Circle

The Goseck circle, although more clearly astronomically based than Stonehenge, is nowhere near as well known as its younger cousin in the Isles. Part of the reason for this is that the Goseck circle was discovered much later than Stonehenge, having not been uncovered until 2003. Stonehenge, on the other hand, had been known about at least since medieval times, as the first written account we have of the monument dates from 1130 CE. It was likely well known long before this as well, due to the nature of the Salisbury Plain, in which Stonehenge stands. The plain is a chalk plateau, and supports grassland but not a forest, much like the central Asian steppe. This would have made it good for grazing, and as good for astronomical monuments. One of the biggest complaints of astronomers through history has been obscuration of the horizon or other parts of the sky, by trees or other objects. However, forest is not thick on the Salisbury Plain. And thus, while sites like Tikal were covered in overgrowth and swallowed up by the lush rainforest of the southern Maya lowlands, Stonehenge likely would have remained visible to travelers on the Salisbury Plain from the time it was built to today.

[4]Pearson (2012).

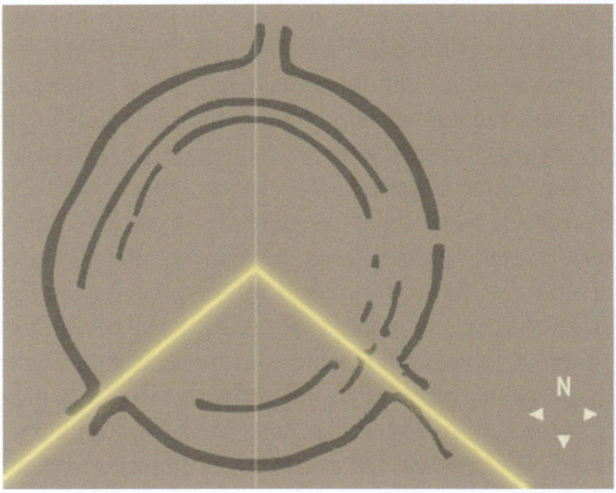

Fig. 8.2 Illustration of the Goseck circle site—a neolithic period solar observatory, aligned primarily to the December solstice. The lines in the photo represent the direction of sunrise and sunset on the solstice

The Goseck circle, unlike Stonehenge (and more like Tikal, though its forest is not as thick), *does* lie in a forested area, and overgrowth eventually covered this ancient site. In addition, the Goseck circle was not constructed of massive pieces of rock as Stonehenge was, and so it would not have been as obvious or apparent that anything of significance ever existed in its location, unlike a site like Tikal, which remained, even though overgrown, a massive city clearly distinct from the region surrounding it. The Goseck Circle was also a much smaller construction, and much of it was earthen, which would have made it very difficult to discern from the ground. It required the age of flight to get the first hint that something of interest lay beneath the cleared field in the Burgenlandkreis district of the German state of Sachsen-Anhalt (once a part of the German Democratic Republic, better known as "East Germany", during the Cold War years). From the air, people were able to see the rough outline of a circle, and recognized that this was unlikely to be a natural feature (Fig. 8.2).

Upon excavation in the opening years of this century, archaeologists discovered multiple layers of circles, similar to what is found at Stonehenge, and even more interestingly, three openings in the outer layer, corresponding to the meridian in the north, and on the southern side two openings aligned with the sunrise and sunset, respectively, on the winter solstice.[5] As is the case at Stonehenge, the alignments will not work correctly today—that is, if one stands at the center of either of the circles at the solstice, one will not see the sun rise through the openings in the wall. But this is due to precession of the equinoxes (discussed previously). A full "wobble" of the earth from one point back to its initial position takes roughly 26,000 years. Thus, the positions of celestial objects today, including the sun, are not the same as they would have been in the ancient world. Archaeologists also determined that the inner layers of the circle were likely lined with wooden posts,

[5]Bertemes et al., (2004).

not wholly unlike the stone posts at Stonehenge. The wooden posts in this case were likely completely closed as a wall around the entire perimeter except for the three openings. The outermost layers of the circle were earthen mounds, akin to what we find in various places in Southeastern North America, in the Mississippian region and its related cultures.

The antiquity of the Goseck Circle is enormous. It was created likely in the 49th century BCE, predating Stonehenge by more than 2000 years. It is one example (the oldest known) of this kind of solar observatory in Europe. Apparently the peoples populating the continent during the final years of the Neolithic era were concerned with solar placement, especially with determining the winter solstice. As we have seen in previous chapters, there are numerous ways of determining the solstices. Although the circle method as used at Goseck is one, the gnomonic method used in China has the advantage of being useful over large swaths of time given precession, because although the maximum and minimum shadows cast by the gnomon will change over thousands of years, there will always be a clearly definable maximum and minimum a close observer will determine. With the circle method of Goseck, Stonehenge, and SunWatch, when there is enough precessional change, the structure becomes useless, and either has to be rebuilt or renovated to align properly. Of course, it is likely that the builders of these sites did not build them with such distant points in the future in mind, and probably had no intention (maybe even in the case of Stonehenge, if it was indeed a non-functional monument) of using the structures for such a long period of time that precession would play any role whatsoever. These early people may not have been aware of the phenomenon of precession—it is impossible to say—though if it is the case that people had been building sites with observatorial function on the spot for hundreds or thousands of years, based on passed down alignment knowledge, someone would have eventually noticed at least that things were not aligning as they were supposed to.

Newgrange

Another enigmatic site of "prehistoric" Europe is the Newgrange mound, in the Boyne river valley of County Meath in the northern part of the Republic of Ireland. This monument dates to about the midpoint between the building of Goseck Circle and that of Stonehenge, around 3200 BCE. Like Stonehenge, much of the monument is made of rock, but in a very different way and in a different layout. The builders of the Newgrange mound constructed layers alternating between earth and stones, and on the outside of the mound, there is a final layer of stone enclosing the entire monument. There is one opening in the stone wall of the monument, which leads to a small chamber within. The opening is aligned, similarly to the Goseck Circle, to the sun at the winter solstice. One major difference between Goseck and Newgrange is that Newgrange has a single opening, rather than the two (with one additional for the meridian) of Goseck. This single opening at Newgrange is aligned

Fig. 8.3 The mound of Newgrange, in County Meath, Ireland, is aligned to the winter solstice similarly to the Goseck circle, but is constructed so as to allow a ray of light into a chamber in the front of the structure on the days surrounding the winter solstice (*Attribution* Jimmy Harris)

with *sunrise* on the winter solstice (whereas at Goseck sunrise and sunset on the solstice were determined). At sunrise on the solstice, the light of the sun shines into the opening and the chamber within the mound, forming a line all the way down the narrow hallway within and to the wall behind. This effect only happens for a few days before and after the winter solstice, and for the rest of the year, this chamber is dark (Fig. 8.3).

Like the Stonehenge site, there are a number of related structures near Newgrange, and also (possibly) like the Stonehenge site, Newgrange was a burial place, probably for high status members of society. As with so many cultures in the Americas, burial mounds could play an astronomical role as well. It was rarely the case that a structure as large and requiring as intensive labor as a Newgrange or Monk's Mound was meant to have only a single purpose. It's also the case that structures are often *repurposed* during their lifetime. This is a phenomenon that still happens today. A building erected to house a post office is later used as a pizza store when the post office shuts down. I knew an excellent place in Dayton, Ohio that houses an art collective, but which was originally built as the home of a taxi company. You can still see the dispatcher's office with its window and the garage in back where part of the fleet was stored. The longer a structure exists, the more likely such repurposing will happen, as the concerns and interests of people change, especially over hundreds of even thousands of years. Thus, it is plausible that Newgrange began its existence as a burial mound primarily, as in the case of Monk's Mound at Cahokia. And indeed, archaologists have found evidence of burials at the earliest layers of Newgrange, along with jewelery and other ritual items that could have been grave goods buried with people—a common practice among various world cultures.

The stone circle on the outer layer of Newgrange is a later addition to the original mound, and may have to do with its repurposing. The large decorative stones outside the entrance to the chamber within the mound seem to suggest that the place had become some kind of important ritual site, which is par for the course for tombs. The alignment of the chamber and entrance with the winter solstice sunset suggests a link between astronomical phenomena and the dead. There are many possible explanations for this. Perhaps these ancient people associated the winter solstice with death and rebirth. The time of the solstice is the darkest part of the year, and all life slows or comes to a stop. The leaves have fallen, the insects have disappeared, and the cold descends. The solstice itself signals a coming return to life. The sun begins to move on its path higher into the sky. The months slowly become warmer, and life begins to return to the land. It signals the coming end of winter, and the return of growth and life of the spring and summer. The associations between this process and that of death are obvious. Individuals age, decline, and die—the coming of winter. But in this death and what they pass on, there is renewed life. The next generation grows and thrives, and this process of life continues. There are more than a few parallels between the cycle of the seasons and the life cycle and the perpetuation of species. Newgrange and other similar sites may be designed to represent in a clear and tangible way this fundamental connection between humanity and nature.

As with many other ancient sites, there have been many outlandish theories put forward as to the purpose or purposes of Newgrange. One of them holds that the monument commemorates or serves as a map of the constellation Cygnus.[6] Part of the evidence given for this is that the inner chamber when its plan is viewed from above looks like Cygnus, and stories about the mythology of the area having to do with swans (the Cygnus constellation is the swan). This is highly implausible for a number of reasons, not the least of which being that it would be near miraculous if the Neolithic people of Ireland who built the monument recognized constellations that originated with the Greeks or Babylonians or their Indo-European ancestors thousands of miles away. Even if they took the bright stars of Cygnus as manifesting a constellation, why would they see it as a swan in the way the Greeks did? We don't find this in other places in the world—why should we find it here?

As with many other sites, there is a desire for mystery and the unknown to yield greater complexity and intricate plan and perceived significance than it actually does. Often times (perhaps most often) mystery covers the normal, everyday, and simple. A mysterious object unearthed in a tomb whose purpose we don't understand turns out to be a farming hoe, rather than a complex tool for calculating the distances between galaxies that shows some early civilization was more advanced than any other of the time. We often look for the spectacular and overlook the normal, and far more likely. But Newgrange and monuments like it show us that the more likely and normal interpretations are in no way boring or unimportant. Indeed, the real turns out to be far more profound and thought provoking than the fanciful, when we reflect on its meaning.

---

[6]Murphy and Moore (2006, 136-159).

The possibility I mention above of the perceived connection between the seasons and the cycle of life commemorated in Newgrange is more likely than farfetched theories like the Cygnus connection. But this simple point, illustrated in a single small pillar of light entering the dark chamber of the tomb on the winter solstice, may be one of the most profoundly spiritual experiences a human being can have. I have never been to Newgrange (although I would one day like to go), but I had an experience similar to the one I imagine the builders of Newgrange were aiming to cultivate, at the tomb of the Mughal emperor Akbar in northern India, near Agra (the city of the Taj Mahal). In the deep part of the tomb, at Akbar's real resting place (there is a recreation of the tomb higher up in the complex), there is a small opening in the wall that allows in only a small ray of light from the sun. The tomb is completely dark but for this ray of light. In the deep gravely darkness, the musty smell of millions of years old earth and the feeling of cold around you, the ray of light represents life, continuity, and eternity. In the moment of experiencing it one feels part of a larger entity beyond oneself. One feels part of humankind, part of the life of the universe itself. It is as if one comes to recognize that though the individual dies, we are all part of that which remains.

Such an experience, which is unforgettable and causes us to reflect on our own meaning, our community, our origins in nature, and the ultimate place of humanity in the cosmos, is far more meaningful and significant than wild and complicated uses attributed to sites like Newgrange. In it, and other sites, we can see just how important the simple recognition of the connection between humanity and the sky can be. It is enough.

Astronomy in Ancient Greece

Many contemporary sources like to trace the origins of the Western tradition back to ancient Greece. While there is as much, or perhaps more, reason to see the cultures of the Germanic, Celtic, and Slavic people (among others), as foundational to the culture of Europe, the long accepted narrative takes the Greeks to play the central role. It is not a completely unjustified story. The Greeks influenced the Romans, who turned to Greek culture as a way of bringing a kind of cultivation to their rustic and warlike society, in which arts, sciences, philosophy, and other things the Greeks prized had never previously gained a foothold. It was after conquering much of the region that the Romans decided they needed to have high culture, to which they looked to the Greeks, well known in the Mediterranean world as cultural scions. We see this even in the Roman epics, such as the most famous, the *Aeneid*, in which Virgil attempts to justify the Roman cooption of Greek culture by making the Romans directly related to the Greeks, through the means of the Greek warrior Aeneas who comes to Rome to found the great city that will one day conquer the Mediterranean world. The Romans, of course, went on to dominate much of Europe, and left much of their legacy in its history, through both their language (versions of which, the Romance languages, are spoken throughout Europe, and even the non-Romance languages such as the Germanic and Slavic languages were heavily influenced by Latin[7]), and in their culture, much of which was modeled after that of the Greeks.

The beginnings of astronomical thought in ancient Greece date back much further than the beginnings of philosophy, but it is in the theories of the philosophers that we find the most developed accounts of the operation of the cosmos. Cosmology, in the ancient Greek tradition, had an important role to play in astronomical thought, including in observation, and it would retain this central role all the way through European history and even into the modern period. As I explain in the next section, this is still with us today.

The central idea of ancient Greek philosophy, in explaining much of what happens in later European astronomy, and lead to much of its tortuous and difficult development, was the notion that the heavens, beyond the meteorological realms close to us, are unchanging, eternal, and perfect. This led to numerous problems in later Western astronomical thought, all of which came to a head by the time of Tycho Brahe in the 16th century.

According to Aristotle, the "first scientist" of the Western tradition, it is impossible that something that admits of the ability to be destroyed is not at some time or other in actuality destroyed.[8] Thus we see on earth that human beings grow old and die, buildings fall down, trees rot and fall. There can be nothing mortal that does not die. As the presocratic philosopher Heraclitus long before pointed out, all things we observe on earth are changing. One cannot step into the same river twice. That is, the collection of water that one steps into passes on toward the sea, so when one steps into the river again, one steps into a different body of water. All things we observe seem to follow this dictum, even the most slowly changing things, such as the very earth we inhabit. While a single individual may not see geological changes in a lifetime, over many generations a people may notice that a hill that once stood tall in the time of their ancestors no longer does, or that a lake that covered the ground and created an island, such as Lake Texcoco which once covered much of the valley of Mexico and surrounded the once island city of Tenochtitlan, burns away because of human intervention and natural causes. Or the Boston Harbor, which during the revolutionary period covered much more of what is now the city than it does today.

The one exception to this seeming constant change is the sky—that is, the sky beyond the atmosphere, the realm of the moon, planets, and stars. Generation after generation, for as long as human history extends, the moon, planets, and stars continue their regular paths through the sky. It is this regularity that allowed peoples such as the Maya, Chinese, and Indians to predict with such accuracy the occurrence of events such as solar and lunar eclipses, conjunctions of planets, heliacal risings, and other regular motions of the heavens. It is for this reason that a people

---

[7]All of which are Indo-European languages, related to languages such as Sanskrit and its derivates and Persian (Farsi).

[8]Aristotle mentions this in part of an argument to show the eternality of the cosmos in Book I of *De Caelo*. Kukkonen (2002, 155-158) discusses commentarial reactions to this position.

can know the position of the sun along the background of the stars for each time of the year, as this works regularly. Of course, there are exceptions here as well, but even these exceptions exhibit a remarkable regularity. Precession of the equinoxes, or the "wobble" of the earth due to the influence of the moon, causes slight shifts in the position of the earth relative to the background stars, a slow motion that unfolds over thousands of years. Some excellent observers, such as the Chinese, recognized this motion, while others failed to.[9] Nonetheless, the regularities of the motion of the heavens is in marked contrast to the myriad changes below.

Aristotle noticed this as well, and theorized that the lack of change in the skies beyond the level of the atmosphere, or the closest layer of the heavens to us, was due to the distance of the outer spheres from us. Aristotle saw this distance as coinciding with the purer, and ultimately eternal motion, uncreated and undestroyed. (*De Caelo* I) The purer material moves outward while the more rough moves inward, and this explains the appearance of greater change in realms closer to us and little or no change in those outer spheres. The idea of the "perfection of the spheres" as it is understood in later Western astronomical though, mainly based in the work of Ptolemy (about whom more later) originates then in Aristotle's considerations, even though he himself did not put it in quite that way. Aristotle was not the first thinker in ancient Greek thought to posit the perfection of the spheres and the unchangability of the heavens, nor was he the most influential in the construction of the theory of the spheres, even though he is often given credit for both.

The idea of spheres or rings to which the planets and the stars are fixed has a deep history in ancient Greek thought, and was well known by the time of Plato and Aristotle. The presocratics discussed this idea, and it took on greater theoretical clarity by the time of the early philosophers. Plato, in his *Timaeus*, discusses the creation and the motions of the cosmos, including the planets (a term we get from the greek *planetes*, "wandering stars"). While the term itself was based on the idea that the planets were defined by their irregular movement through the night sky, unlike the more orderly stars, Plato argued that the motions of the planets were also orderly and simply followed a different regularity than that of the stars. Plato identifies both the planets and the stars with divinities, arguing that these objects must have souls, which in part explains their motion (*Timaeus* 38e). But perfect motion, for both Plato and Aristotle, was *circular* motion. The moon, planets, and stars, as distant, unchanging, and perfect, move in circles around the earth. Indeed,

---

[9]The Chinese knew of precession at least as early as 330 CE, and likely before. The shift in the solstice position against the background of the stars due to precession was noticed at least early as the 2nd century BCE, but Yu Xi's work clearly indicates knowledge of precession. Sun and Kistemaker write: "Yu Xi discovered a constant called *shui cha*, which is the difference between the lengths of the sidereal and tropical year. This he explains with the annual westward shift of the winter solstice along the ecliptic (Sun and Kistemaker.: 112). In the *Han shu*, there is a documented account of the effect of precession, but it is attributed to carelessness by court astronomers (Pankenier 2013: 249).

it is in part this constant circular motion that proves to Plato and Aristotle that the celestial bodies are indeed perfect and unchanging. And because of this, it should be possible to predict their movement and position at any given time.

The theory of rotating spheres was developed most fully by a rough contemporary of Plato and Aristotle, Eudoxus,[10] who did more than perhaps any other ancient Greek thinker to develop what became the Western tradition of astronomy. Eudoxus hailed from the eastern part of the Greek world, in what is today western Turkey, along the eastern part of the Aegean Sea. His hometown, Cnidus, was south of the home of another luminary of ancient Greek mathematics, Thales of Miletus (and the later Milesian "presocratics"). In his youth, he studied under various astronomers and philosophers, traveling as widely as Athens and Egypt to gain a greater understanding of the world and especially the sky. Rather than being the student of any particular scholar, Eudoxus appears as a wandering savant, who learned from anyone he could, and developed his astronomical systems on his own with the help of the wide learning he accrued through the years. After becoming himself an accomplished and well-known thinker, Eudoxus taught a number of students, and a "school" formed around him. He was later to have further contact with the forebears of Plato in Athens, including Aristotle, in his discussions with members of the Academy on his second stay in Athens. These discussions likely took the form mainly of debates, as there were serious disagreements between the school of Eudoxus and the Platonists, including the former's rejection of metaphysics based in invisible realms like that of the Forms, as well as his view that pleasure is the highest good for humans (a position Aristotle presents a lengthy argument against in Book 10 of his *Nicomachean Ethics*). We are forced to take Aristotle's word for this, as his work is the only source of the claim concerning Eudoxus' hedonistic views. We might imagine, however, that Aristotle set up a straw-man, and thus should not discount the possibility that Eudoxus' view was quite different than what Aristotle claims.

One very important feature of Eudoxus is that he revolutionized the concept of the celestial sphere, an important innovation that existed at least back to the time of Parmenides in the 5th century BCE, but also was in use in ancient China, as discussed above in the chapter on ancient Chinese astronomy. Eudoxus also applied mathematical techniques to understanding celestial motions, based on the motion of spheres.[11]

Eudoxus' understanding of the motion of celestial bodies was to become massively influential, in part through its adoption by Aristotle, who would become, in the Western world, both in Europe and the Middle East, the driving force behind intellectual development in the medieval age, and thus the foundational thinker in

---

[10]Eudoxus is said to have studied under Plato for some time (from the account of Diogenes Laertius), as well as the sophists, when he came to Athens in his 20s, before moving on to learn from astronomers in Egypt, but it is unclear to what extent he could be considered Plato's "student". Eudoxus returned to Athens later in his career (Dicks 1970, 151-152).

[11]A detailed account of Eudoxus' astronomical work can be found in Dicks (1970, 151-189).

the development of scientific thought in both areas. Eudoxus' use of the sphere and his conception of the celestial bodies moving consistently with the surface of the sphere led to his (and future) considerations of planetary (and other celestial) motion as circular motion. The notion of the celestial sphere did not originate with Eudoxus—Parmenides and earlier thinkers also had the concept of the universe as spherical, but his system of explanation of celestial motions by the use of a theory positing numerous rotating spheres was a unique development, and one that was to have a long history in Western thought.

Eudoxus had to explain a number of different kinds of motion, that seemed inconsistent with one another. If if were the case that the earth rotated in the plane of the ecliptic, and the moon orbited parallel to the plane of the ecliptic, Eudoxus would have had a much easier task. He would have required fewer spheres—some for diurnal motion, and others to make sense of the motion of the sun, moon, and planets relative to the stars—in the case of the moon and planets, due to their own orbits, and in the case of the sun, due to earth's orbit. Eudoxus did not have this luxury. He had to explain not only irregular diurnal motion, in terms of the motion of the sun and that of the stars, but also the motion of the planets offset from the celestial equator, in general following the motions of the sun (through the ecliptic), and that of the moon around the earth, which is also not in the plane either of the celestial equator or the ecliptic, as the moon's orbit is inclined 5° from the ecliptic, which makes it variably as much as 28.5° elevated to the celestial equator. It can be maximum this many degrees above or below the equator, and anywhere in between at different times.

In order to make sense of all of these complex and sometimes seemingly irregular motions, Eudoxus devised a system of multiple concentric spheres, involving a number of elements. The Earth, naturally, was at the center of this system of spheres, and the motion had to be explained as circular motion based on connection to spheres in outer layers of ever increasing distance. The motion of the celestial bodies, Eudoxus noticed, were not uniform in a number of ways. There was the fact that they moved in different paths, and also that there were variable speeds of their motion (we know now that this is *apparent*, rather than actual speed). One of the main problems Eudoxus needed to solve was that while the sun and planets move through the year, the ecliptic upon which they appear to travel also changes its path, inclining to a greater or lesser degree from the celestial equator. In specific, it moves between 23.5° above it, on the June solstice, and 23.5° below it, on the December solstice—these should be flipped for the southern hemisphere, of course, but Eudoxus knew nothing of that. Eudoxus accounted for this by theorizing that diurnal motion is due to one sphere, turning around the axis of the earth, along the celestial equator. Outside of this sphere, which it turns out is actually a collection of spheres, as each planet, as well as the sun and moon, has its own collection of spheres associated with it, for reasons I discuss below, there is another sphere, attached to the poles of the inner sphere, and this second layer moves the whole structure, starting from the equinox points (when the ecliptic is 90° to the Earth's pole, which would have been determined by the pole star or the pole star region, when there was no single star at the pole). This sphere wobbles

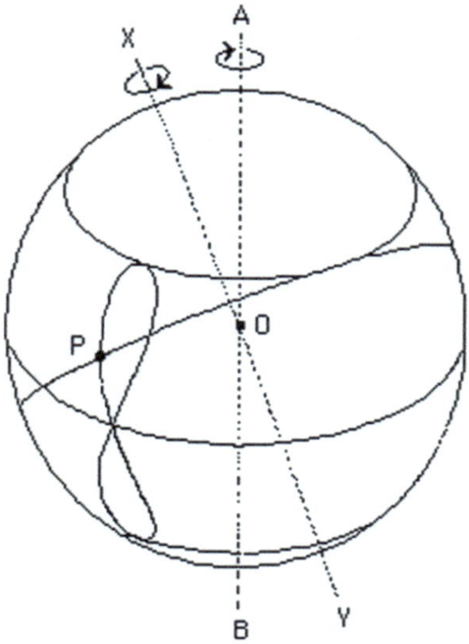

Fig. 8.4 The figure-8 shape of the *hippopede* that Eudoxus used to explain retrograde motion was due, he argued, to the counter-rotation of differently situated spheres, as shown in this illustration

back and forth, 23.5° on either side from its connection at the poles, making one back and forth trip each year (that is, from equinox back to that same equinox the following year) (Fig. 8.4).

So much for the annual motion of the ecliptic and the diurnal motion of the sun and stars. But what of the motion of the sun against the background of the stars—that is, its annual progression through the zodiac (*zoidiakos*), its motions against the sphere of the stars? And what of the motion of the moon? And perhaps the most difficult question Eudoxus faced—what can we say about the strange motion of the planets? To explain solar and lunar motion, Eudoxus added yet more spheres, tethered to but moving independently from the others. There were three spheres, according to Eudoxus, responsible for the motion of the moon, and three also for that of the sun. One of the spheres accounted for diurnal motion, rotating once every day. Each of the celestial bodies and the sphere of the stars had such a sphere rotating daily. These had to be distinguished from one another, rather than attributed to the same sphere, because these bodies are observed to move independently of one another. The second sphere associated with the moon was also one the other celestial bodies had analogues of, as it accounted for motion through the zodiac (which for the sun was annual, but for the moon monthly, as the sidereal period of the moon—the time it takes for the moon to reach the same point against the stars—is 27.3 days). A third sphere of the moon accounted for the motion of the moon above and below the ecliptic, as the moon's orbit is elevated 5° from the ecliptic. The motion of this sphere would have been similar to the "wobbling" of the sphere determining annual motion of the plane of the ecliptic, only relative to the ecliptic rather than the celestial equator.

The three spheres associated with the sun worked similarly in the case of the first two spheres—one accounting for diurnal motion and the second for zodiacal motion. The third sphere was meant to account for a strange motion that was actually based in a mistake—the idea that the latitude of the sun changes, because of perceived differences in the places of sunrise and sunset on the solstices. The issue of the motion of the planets was the most difficult puzzle to solve. There is one important reason for this. The planets, even though they appear generally to move in a single direction relative to the background stars, sometimes appear to pause in this motion, move in the opposite direction for some time, and then reverse themselves and move in their original paths. It is as if they are making an awkward loop in their path. We know today that this phenomenon, what we call "retrograde motion," is due to the motion of the earth in its orbit relative to the planets in theirs (about which more below). But the ancient Greeks did not know this, nor did anyone in the Western world until about two millennia later.

Eudoxus' strategy for making sense of planetary motion was to add a fourth sphere for each of the planets to account for retrograde motion. Two of the spheres were similar to the first two spheres for the moon and sun—that is, accounting for diurnal motion and, as with the sun, motion in the ecliptic. The second sphere moved with different speeds for different planets, according to the sidereal period of each of the planets. The third sphere accounted for the *synodic* period of each of the planets, and the fourth sphere rotated in the opposite direction of the third, at the same speed (for each planet), inclined at an angle to the third sphere. This created a shape like an 8 or infinity sign called a "hippopede", based on the movement of the third and fourth spheres. With the additional motion of the second sphere, this hippopede would be moved in the direction of the planet's orbital path, and retrograde motion would happen whenever the motion of the hippopede ran against that of the second sphere, that is, opposite the direction of the motion of the second sphere. Variations of this complicated solution to the problem of retrograde motion would last through most of later Western history of astronomy, culminating in the theory of planetary epicycles, which require a movement away from the celestial sphere conception of planetary motion, but nonetheless shares key similarities with Eudoxus' theory. The notion of circular motion in general, enshrined by early luminaries such as Aristotle (who took most of his positions on astronomy from Eudoxus), would remain a fundamental tenet of Western astronomy, along with one further idea that would die hard in modern history, the notion of the fundamental perfection and immutability of the spheres.

Why did the ancient Greeks accept such an idea, which should have seemed to them fundamentally flawed? After all, the Greeks must have noticed comets and other non-regular celestial phenomena, such as supernovae. Plato and Aristotle both give us some hint of why this should be. Circular motion, according to both, is perfect and unending. Why might they think this? Circular motion does not have a natural stopping place, even without being an infinitely long line. That is, with any circle, once a motion started at one point reaches the end (the point just before the first point), it is still on the path of motion, moving back to and through the first point and around the circle. There is no natural stopping point of a circular path, no matter how large or small the circle is, which is fundamentally different from travel in a line, which either has finite and stopped points on either side, or continues infinitely in one or both directions—though notice that travel along such a path will involve traveling through a new point endlessly. Aristotle argues that time itself is akin to the infinite line reaching in both directions.

According to Aristotle, those actions and objects within the closest spheres to us, that of the sky and the earth itself, are not perfect (*De Caelo* 1). It is obvious to the senses that things in this realm undergo drastic and constant changes, and anything that changes in this way cannot be perfect, because change requires unactualized potentiality—having the potential to be something a thing is not currently. Perfect and eternal things are fully actual and not potential. This view, applied to the cosmos, was to have enormous influence even until the modern era.

The final ancient Greek thinker worthy of discussion here is Hipparchus, who was from, as many of the great mathematicians and astronomers before him, the eastern part of the Greek world, in what is today Turkey. He came from the city of Nicaea, well known today as the city that hosted the first Ecumenical Council of the Christian church, convened by the Roman emperor Constantine (who created the orthodoxizing movement we know know as "Christianity") which resulted in the "Nicene Creed", the central dogma of Christian doctrine.

As mentioned above, no writings of Eudoxus remain, and our knowledge of his theories comes mainly from two sources, Hipparchus and Aristotle. Hipparchus, of course, was a major mathematician and astronomer in his own right. Many attribute to him the invention of trigonometry, or at least the systematization and use of it to solve important problems. His knowledge of spherical trigonometry was certainly advanced, and it enabled him to create (or possibly *invent*) one of the most widely used astronomical tools in Western and Middle Eastern astronomical history—the astrolabe.

Hipparchus is also known for his discovery of precession of the equinoxes (a phenomenon discussed in depth above). Although he may have been the first person to give an account of the phenomenon, he was almost certainly not the first person to notice it, and it may have been recognized by people as early as the Neolithic periods discussed above. Any culture that had a regular and sustained routine of astronomical observation over many generations would begin to recognize that alignments and positions altered slowly as time went on. It is difficult to imagine that constant skywatchers like the builders of Stonehenge or the ancient Chinese did not recognize precession. Nonetheless, Hipparchus gives us an explanation of precession, and from his observations and comparison with the observations of earlier astronomers, he found a number of consistent differences, which by itself would not have said much. It was possible, for example, that Hipparchus' observations were more (or less) precise than those of earlier astronomers, or that his system of coordinates was somehow different. It would be an enormous coincidence that other than the slight errors all positions matched, but possible nonetheless. But Hipparchus also observed a small but perceptible difference between the sidereal and tropical years, which showed that the equinox points are not exactly in the same position against the background stars every 365 days. This movement of the equinox points was thus called the precession of the equinoxes. We know today that this is caused by the "wobble" of the earth due to the gravitational influence of the moon. Precession of the equinoxes, then, refers to the process Hipparchus discussed, not its cause (which is the gravitational pull of the sun and moon on the equator, where the earth bulges due to centrifugal forces along this plane due to the motion around the axis). He determined this precession to be a bit more than 1° of arc per century. The currently accepted value of precession is 1° every 71.6 years.

Hipparchus was also able to determine a number of important facts about the moon, including that its orbit is tilted 5° to the ecliptic, and also (perhaps more impressively) a fairly accurate determination of the distance from the earth to the moon. Hipparchus was able to do this by using the phenomenon of parallax shift (discussed in Chap. 2). In the case of determining the parallax shift of the moon, if we know the speed of the moon's orbit, and thus how much of that orbit will have been traversed through a day of moon visibility, then we can factor this into the observation of the moon against the backdrop of stars on a complete night (as long as we can see the moon), observing at the beginning and at the end. Factoring in the moon's motion, we can calculate where the moon would have been had it not been traveling, given the second observation, and compare this with the moon's location in the first observation. Due to parallax shift we should notice that they are slightly skewed. Why this should be is straightforward. Since through a night the earth is rotating around its axis, the angle from which we view the moon on the second observation will be different than that on which we view the moon on the first observation, which, factored in with the movement of the moon, gives us an indication of the parallax shift. Using this information, a mathematician armed with trigonometry, as Hipparchus was, could relatively easily determine the distance of the moon (or any other object for which parallax could be measured). If we know the radius of the earth—that is, the distance from a point on the earth's surface to the center of the earth—then we can know the distance to the moon given a few more elements. If we make a first observation of the moon and note its location, then wait 6 h and do the same, earth will have completed 1/4th of its rotation in this time, and the angle of the radius between one's point 6 h later and the center of the earth with the initial radius line will be 90°. The second observation will give us a different angle between the moon, our point of observation, and the center of the earth. This new angle shows us that the viewing angle from earth has moved some certain amount in a change of one earth radius difference in space. From this, we can calculate the lunar distance.

But how do we find this initial radius, for which some knowledge of the circumference of the earth is necessary? Fortunately Hipparchus had some earlier calculations to work with (even though they were flawed). Eratosthanes, many years before, had calculated the circumference of the earth by comparing angles cast by a gnomon at different places on the earth. He had heard that the summer solstice sun stood directly at the zenith on noon at the summer solstice at Syene, in lower Egypt, and so took a measurement of the angle created by the shadow cast by the sun at noon on the summer solstice at Alexandria, a distance of 5000 "stadia" north of Syene (Eratosthanes used the average length of a stadium as a unit of measurement. The angle created between the gnomon and the edge of the shadow downward (found by knowing the length of the gnomon and that of the shadow on the ground, and the angle of 90° between these two) will show you how much of a circle is represented by the distance between Alexandria and Syrene. Eratosthanes determined that the angle was 1/50th of a circle, or 365/50 = 7.3°. If 7.3° corresponded to 5000 stadia, then the entire distance around the world must be 5000 * 50 = 250,000 stadia. Eratosthanes later revised this distance downward. Nonetheless, if we have an idea of the circumference of the earth, then we can calculate its radius.

Hipparchus' contributions were certainly great (some say that Ptolemy's thought was mainly a reiteration of that of Hipparchus, although there is little evidence for this), and he was certainly the most influential astronomer for hundreds of years after his time. Today little of his own work survives, and his work is known mainly through the work of the man who would both supersede him and become the most influential astronomer in Western history until the time of Isaac Newton. The cosmology of the spheres is and the astronomy of the ancient, medieval, and modern West as more closely linked with him than any other figure—more than anyone else, Ptolemy dominated astronomical thought. For a large swath of history, in Europe and the Middle East, Ptolemy's work *was* astronomy—the first and last word on the subject.

Ptolemy and the Spheres

Though the influence of Aristotle is undeniable in Western astronomy, the main reason Aristotle is so influential is due to the massive influence of another figure, arguably the most influential in early Western history. The astronomical thought of the Egyptian polymath Claudius Ptolemy (better known as simply Ptolemy) dominated Western and Middle Eastern thought about the cosmos for more than 1000 years. His major astronomical work, the *Almagest*, was the most influential astronomical text in Western history, if measured by the standard of use over time and place, and even one of the most influential texts of *any* kind in Western history. Its theory of the cosmos was accepted in some form or other by almost everyone in the Western and Islamic world up at least until the time of Johannes Kepler. Even Copernicus, who radically altered the Ptolemaic system to out put the sun, rather than the earth, at the center of the cosmic system of rotations, retained much of the Ptolemaic system. And later astronomers such as Tycho Brahe devised theories of celestial motions even more like that of Ptolemy. The Tychonian system, developed in the late 16th century, still operated on the same assumptions and used the same ideas of motion and change as that of the Ptolemaic system, devised almost 1500 years before. It would take the careful and insightful mathematical theory of Kepler, based on the painstakingly intricate observations of Tycho Brahe, to finally put the Ptolemaic system to rest. And even then, the ideas of Ptolemy, and Aristotle before him, had influence in Western ways of thinking about astronomy. The Newtonian "revolution" was only a partial revolution, just like all revolutions. When the American colonials rebelled against British rule, they established their own nation, but they didn't do away with British customs, language, and ways of thinking about the world. Ptolemy and Aristotle would continue to have major influence throughout the rest of Western astronomical history even until contemporary time.

Most of Ptolemy's life was in the second century CE, and he inhabited what had been one of the greatest centers of learning in the world, the majestic city of Alexandria in Egypt. There is actually some disagreement as to where Ptolemy was *from*, but we know from the *Almagest* that he made observations in Alexandria, so he is generally associated with this city, although he may have hailed from elsewhere. Some evidence shows he may have been from the nearby town of Canopus. By Ptolemy's time, the city of Alexandria was the cultural and intellectual center of the Hellenistic world. Alexander the Great founded the city, which had been built in a relatively remote area in Egypt, in 331 BCE. By about 100 years after this, Alexandria had become the foremost center in the world of Hellenistic learning. Even before this, Egypt was known for its learning, and we saw earlier that Eudoxus continued his astronomical training after visiting Athens by learning from astronomers in Egypt. The Romans officially annexed the city in 80 BCE (though they had influence in it long before), and it remained under their control until the closing years of the Roman Empire. Ptolemy's Alexandria was still a place amenable to learning, with its still famous library containing much of the learning of the ancient world (and which still likely stood in Ptolemy's time, before its destruction, though there are different accounts to just when this was, and how extensive) and in which it would not have been strange to find people like Ptolemy, the kind of multi-faceted "renaissance man" so rarely seen today. Ptolemy was not only a mathematician and astronomer, but also a philosopher, geographer, and poet.

There is one feature of Ptolemy's thought that was quite common through most of the history of Western thought, but tends to be written out of or at least neglected in histories of Western astronomy. In addition to his most well known work, the *Almagest* mentioned above, Ptolemy also wrote works on geography and astrology that were as well known for a large time as the *Almagest*. Ultimately it was his work on astronomy that remained influential for the longest time, and people used his astronomical material long after his geography had been superseded and his astrological concerns redirected (often in different astrological systems like that of Kepler). The distinction we make today between astrology and astronomy was one not recognized in many cultures and for much of history, and most astronomers in the Western tradition had what we might call astrological interests that were combined with their work we would deem more scientific and thus astronomical.

The text *Almagest* was given this name not by Ptolemy himself, but, as might be expected, later Islamic scholars. The original title of Ptolemy's work was "Treatise on Mathematics" (*Mathematike Syntaxis*), which later became known as the "Great Treatise", and the Arabs translated the "Great" from the Greek *megiste* as *al-majisti*, which became, when the text was introduced to the West as so many other ancient Greek texts through the Islamic world, *Almagest*. The fact that the text comes down to us associated with this title shows both how influential this text was in the Islamic world, and how much the Western world relied on the texts and learning of the Islamic world, which as we saw in the previous section, had inherited the knowledge of the Greeks, as they controlled the places such as Alexandria and the eastern Meditteranean that had previously been the center of Hellenistic thought.

In the *Almagest*, Ptolemy presents astronomical thought of the ancient Greeks, as well as develops his own unique system. It is clear that Ptolemy is most highly influenced by Aristotle, as the opening of the *Almagest* includes an account of Aristotle's view on the spheres, one which we see goes back to Eudoxus, whose account of the operation of the spheres developed earlier views of celestial spheres to a more theoretically advanced point, and began the mathematical (geometric) study of cosmology in Western history. Ptolemy's text gives and develops the accounts of earlier Greek astronomers, including Hipparchus, whose work is mainly known through the *Almagest*. Ptolemy would have had access to sources that are now lost, like many other scholars of his time. There is some disagreement even today about the originality of Ptolemy's work. Some see his role mainly as a compiler and systematizer of earlier knowledge—the theories of thinkers like Plato, Eudoxus, Aristotle, and Hipparchus. Others note original contributions of Ptolemy, in addition to his synthesis of the astronomical thought of these earlier figures, which is a major creative feat in its own right. In addition, Ptolemy did not just list the older systems he gave accounts of in the *Almagest*, he also improved on them, fixing the inconsistencies and flaws in these systems (as much as possible), in an impressive way for one of his era. By any measure, Ptolemy was a very original and brilliant thinker.

The Problems of Ptolemy

Even for all of Ptolemy's innovations and solutions of various problems with earlier astronomical systems such as those of Eudoxus and Hipparchus, there were still a number of discrepancies between the Ptolemaic model and the observed motions of the celestial bodies. These discrepancies would become more and more obvious as observational methods and instruments improved, until, in the 16th century with the revolutionary observational changes of Tycho Brahe, it became clear that Ptolemy's (and ironically Tycho's own) system could not be maintained in the face of the available data.

One aspect of Ptolemy's thought that tends to be neglected today (just as it is in the case of Kepler and other astronomers) is what we tend to call his "astrological" thought. The division between the scientific field of astronomy and the superstitious or mystically-minded astrology is a contemporary one, and reflects our concerns, not those of the ancients (or the moderns, in the case of Kepler). The issue of astrology is an important one if we aim to understand the ways people in the ancient world thought about the skies. We cannot fully understand their thought if we pick from it that which we think sufficiently fits our own concerns or criteria of seriousness and rigor, and ignore the rest. Ptolemy's geometrical understanding of celestial motions, just like Kepler's, was informed by his understanding of the human significance of those motions. This is a view we also see in the other cultures examined in this book. The cleft between accounts of the workings of the universe and the significance of these to human life is relatively new in human history, and may not completely be a good thing. It has, in part, led to the professionalization of astronomy and its decline as a concern of humanity in general. The final frontier of the "sacred" (as we see with ancient sites like Newgrange, Stonehenge, and those mentioned in previous chapters) was the celestial realm, and today that too has been desacralized by our ways of thinking—even though this is still resisted in some ways, as we see in general public television shows where the wonder of the universe is maintained. We have a fragmented consciousness about the cosmos today. While we tout the wonders of the sky, at the same time we attempt to strip it of meaning through our investigations. It is not that the scientific investigation of the cosmos in itself is a bad thing or leads to this, but rather our "scientific" attitude that any human emotion, sense of meaning, or concern must be left out of our investigation—this is what strips the cosmos of the significance it held for so many peoples. And this is what leads us to want to draw a sharp and clear line between (legitimate) astronomy and (illegitimate) astrology, or any other way of understanding the cosmos. We have essentially stripped away any other way of thinking about or investigating the cosmos than that of modern science, with its background assumptions. Think of how many times you've seen courses on astronomy or anything having to do with the universe outside of the earth in university settings that deal with something other than physics. Sometimes you may see a history of astronomy course, but even here, you will not find much mention (if any at all) of the astronomical systems of the Maya, the Vedic astronomers, the builders of Newgrange, or Ptolemaic " astrology". What you will often find is a history shorn of everything but that which is consistent with the ways we today understand the sky.

The "astrology" of Ptolemy's day bears striking resemblance to early Chinese views about the continuities between nature and humanity—an idea that all systems of astrology rely on (I did not use the term "astrology" in connection with Chinese thought on these issues, mainly because this only becomes an issue in contemporary Western thought, in discussions of this very material. The Chinese *tian wen*, which we can translate 'astronomy', was broadly concerned with understanding of the cosmos, in both its operation and its significance to humans). Astrology, in the sense of the patterns of the heavens both signaling phenomena in the human world and having an effect on these phenomena, was taken seriously in the West in Ptolemy's time, although it was dismissed by some in later Roman times. It was finally the rise of Christianity that led to a decline in the influence of astrology (and astronomy in general, as the two were linked). The Christian movement would have had plenty of reason to reject the teachings of astronomers/astrologists, as the stated significance of the heavens in human life and the location of spiritual concern and significance in the cosmos would have been seen as inconsistent with and possibly undermining the Christian position, which was that all spiritual significance is in Christ alone, and not in the world, certainly not in nature. The idea of nature reverence in any sense, let alone worship, was intolerable for the Christians.

Visions of Change—Brahe and Kepler

The notion of the spherical movement (or variations of this) of celestial bodies around the earth and the various attempts from Ptolemy to Tycho Brahe (despite the heliocentric system of Copernicus) was finally fatally undermined by the theory of planetary motions of Johannes Kepler, which was made possible by Tycho Brahe's painstaking and precise observations, themselves made possible by his innovations in observational tools (many of which were anticipated by Chinese tools, which accounted for the accuracy of their observations. Had the Chinese methods been available in the West, Kepler's innovations may have happened over a thousand years earlier). The observation of events such as comets and supernovae undermined the Aristotelian notion of the perfection and thus immutability of the spheres or celestial motions (those of Kepler's time were treated to a very rare occurrence of *two* supernovae visible within 32 years). In addition, the voluminous data compiled by Tycho Brahe ironically spelled doom for his geocentric variation of the Ptolemaic model of planetary motion, as Kepler was able to make better sense of the observed motions on a heliocentric theory following his laws of motion. Kepler's own early acceptance of the Aristotelian view of the perfection of circular motion, and thus the necessity for celestial bodies to conform to this motion, was challenged by Tycho Brahe's data. Ultimately, Kepler could make no sense of the data on a theory of circular motion, and had to abandon such a position to formulate his laws of motion. The fact that Kepler followed the data rather than his pet theory (and one that was by no means just his, but had been accepted by luminaries of astronomy for over a thousand years!) shows a remarkable commitment both to his mentor Tycho Brahe, whose observations he trusted (he knew how stringent and accurate Tycho's methods were, having worked with him), and to the empirical method in general. Astronomers from the earliest periods of Babylonian and Greek astronomy up to Kepler's own time often privileged theory over observation. Many of them recognized that there were discrepancies between their theories and observed motions. Ptolemy, for example, was aware of the problem of the seeming inclination of the planes of planetary orbits, and attempted to account for this in a later work. Hipparchus recognized the inability of his theory to accurately predict lunar positions outside of opposition and conjunction. Part of their failure to revise their theories is due to inability, certainly. But it is also the case that theory took precedence, just as we see in the work of the philosophers Plato and Aristotle, and more or less minor discrepancies between theory and observation were not thought to be as great of a problem as they later were, and now are. With Kepler we see a move away from the commitment to theory first, and toward the very modern idea of being led by the data. Modern science in part, then, was born from features that were not completely dictated by "scientific" concerns—particularly loyalty, trust, and open-mindedness. None of which would be considered particularly scientific today.

Kepler's laws, as we generally encounter them today, in their mathematical and corrected form, were devised by Isaac Newton. But certainly it was Kepler who first formulated the principles behind these laws, creating their initial versions, and their core is in the version of Newton. It is for this reason that we refer to them as "Kepler's Laws of planetary motion" as distinct from "Newton's Laws", which are more general, and actually, as it turns out, are most problematic in the realm of astronomy, as they do not account for relativistic phenomena. Understanding of this would have to wait for over 200 years.

Kepler's work was made possible by the observational data of Tycho Brahe, whose precision instruments and meticulous observational methods made possible better data than had ever before been available in the West. Tycho Brahe had been a major figure in astronomy renowned throughout Europe by the time Kepler joined his staff of observers. A famously tempestuous character, Brahe came from the upper class nobility of Denmark (though his ancestral home, then part of Denmark, is today in Sweden). Among his earlier exploits, which contributed to his legend in later years and after his death, Tycho lost his nose in a duel. He later wore a prosthetic to cover (as much as possible) the injury. He amassed a massive wealth through both his family as well as through the fame he gained in his native country from his advances in observational astronomy. Tycho had from a relatively young age devoted himself to astronomical work, and managed to generate a reputation in it through his skill. But it was with his work on the supernova of 1572 (which would come to be known in history as "Tycho's supernova") that his fame would be solidified. Tycho's 1573 tract on the supernova showed that Aristotelian notion of the "perfection of the spheres" must be incorrect. The "new star" could not be a meteorological phenomenon or something taking place in a sphere close to the earth (more corruptible according to Aristotle in Book I of *De Caelo*). If the "new star" was something closer than the moon, then it should exhibit clear parallax shift. The supernova exhibited no measurable shift, which showed that it had to be part of the "fixed stars". This clearly showed that this distant "sphere" was changeable. Tycho would go on to further demonstrate this in another work on the comet of 1577. His careful observations of the motions and parallax of the comet showed that it was 5 times more distant than the moon.[12]

Tycho's renown eventually earned him a commission from the Danish king, with which he built his famous observatory and residence Uraniborg on the island of Hven. Soon after he would construct another observatory on the island, the partially underground Stjerneborg. The Stjerneborg site still stands today, while Uraniborg is long gone, having been destroyed in 1601. It was only the year before this that Tycho met Johannes Kepler, who he went on to hire to work in his new observatory. Kepler would only work with Tycho Brahe for about a year, as Tycho died in 1601. According to the traditional story, on Tycho's deathbed he bequeathed his observational data, which he carefully guarded in life, to Kepler. He knew Kepler's theoretical genius, and desired his immortality through Kepler. Tycho had already fallen out of favor in his native country, and was concerned he

[12]Heidarzadeh (2008 41-43). This section discusses both Tycho's work on the 1572 supernova and the 1577 comet.

would leave no lasting legacy. He was reported to have told Kepler "let me not seem to have lived in vain." What we do know is that Kepler had access to Tycho's data, and with it he was able to formulate the three laws of planetary motion that would transform Western astronomy.

Perhaps ironically, Kepler's laws contributed to the downfall of Tycho's own cosmological system. Tycho never fully adopted a Copernican heliocentric model, but decided instead to advance a modified version of the Ptolemaic model. Instead of *all* of the planetary bodies and the moon and sun revolving directly around the earth, Tycho's model took the sun and moon to orbit the earth directly, while the other planets orbit the sun. This was in part meant to account for retrograde motion of the planets. With Tycho's data, Kepler was able to devise a theory of planetary motion for more elegant and accurate than Tycho's. We shouldn't judge Tycho harshly for his failed theory, however. Tycho's major contribution was his advances in observation and instrumentation, while Kepler's was his theoretical skill. The result of the combination of these two were the three laws that we today know mainly through Newton's formulation. Kepler's Laws are simply described, and account for all of the observed motions that so many other theories had to construct elaborate concepts like epicycles to make sense of.

Kepler's first and second laws appeared first, part of his 1609 work *Astronomia Nova* ("New Astronomy"), while his third law appeared first in his 1619 work *Harmonices Mundi* ("Harmony of the World"). The laws, though revolutionary, are stated simply. The first law simply holds that, unlike many earlier theorists had claimed, the orbits of the planets are *ellipses*, and the sun serves as one of the foci of each of these elliptical orbits. Ever since the time of the ancient Greeks, circular motion was taken to be the natural motion of the planets, as (so Aristotle argued) circular motion was perfect motion. Kepler's second law holds that an equal area is carved out by the line between a planet and the sun in equal time. Thus, if it takes a planet 10 days to carve out an area of x, it will take another 10 days to carve out an equal area. This would be unsurprising and uninteresting if the orbits of the planets were circular with the sun at center. Because the orbits of the planets are elliptical, this means that the planets will move with *variable speed* around the sun. The planets move more slowly when further from the sun, and more quickly when close to it, consistently with Kepler's Second Law. When the line from the planet to the sun is drawn from the planet's point of aphelion (greatest distance from the sun), it sweeps out a larger area in an equal distance of the planet along the orbit than does the line between the planet and the sun at its point of perihelion (closest distance to the sun). Kepler's third law, also called the "harmonic law", is in some ways the most abstract, and took Kepler longest to determine. It says that the square of the orbital period of a planet is proportional to the cube of the average distance between the planet and the sun throughout its orbit. The third law cannot, however, be applied to objects outside of the solar system. Newton's reformulation of Kepler's third law, informed by his theory of gravitation, made the law generalizable. Newton's modifications were necessary because the *masses* of the two objects involved—the planet and the sun, in this case—are relevant. As Newton discovered, any two objects are orbiting *each other* around their center of mass. Given

Newton's third law of motion, we know that gravitational force exerted on a planet is opposed by equal gravitational force on the sun. In the case of the planets in the solar system, the center of mass between any of the planets and the sun is within the sun itself. Even for the most massive of the planets, Jupiter, the center of mass between the planet and the sun is about at the surface of the sun.

Kepler's formulations of the three laws transformed astronomical thought in the West, and even today astronomical thought in the West is related to the thought of Kepler. There is one fact about Kepler's thought that is often neglected in histories of astronomy, and by contemporary astronomers. Kepler's laws themselves, especially the third law, were grounded in clear philosophical and religious views that many contemporary scientific thinkers would find objectionable. There were certain philosophical assumptions underlying Kepler's laws that were as influential in his astronomical thought as the philosophical assumptions underlying Maya, Indian, or Chinese understandings of the cosmos were on their own. A major assumption of Kepler's, one still entrenched (though in slightly different form) in much of Western thinking today, is that a divine order is manifested through physical objects. This assumption led to a number of different positions—not only Kepler's revered laws, but also his less discussed "polyhedral hypothesis" based on the Platonic notion of perfect shapes created by God.[13] The idea of an order determined by a divine agent was not unique to Kepler, nor was it new. Plato in his *Timaeus* spoke of a "demiurge" who crafted the world using the perfect and eternal forms as its basis (*Timaeus* 28a1–28c9). Kepler inherited a position much like this, and it informed all of his astronomical work, including even a epistemological position that guided his methodology—the distinction between appearance and reality, and the human ability to understand "reality" through discovering the divinely created patterns manifest in material objects.[14]

Kepler's religious and philosophical assumptions play a large role in contemporary "Western" astronomical thought, and failure to recognize this role can lead to a kind of arrogance concerning contemporary understanding of the cosmos. Because we ignore the existence of these philosophical underpinnings, it becomes easy for us to think that our own astronomical thought, that of the contemporary world, born in Europe and the Middle East, is uniquely independent of religious or philosophical assumptions, and for this reason is more "objective" or justified than other systems of astronomical thought, such as those surveyed in this book.

[13]"To the problem of accounting mathematically for the relationship between the planets' distances and their periods, Kepler now added the question of accounting for the number of planets and their particular distances from the sun. He promptly hit upon an explanation for the latter problems. In his "polyhedral hypothesis," he reasoned that God had used the five perfect Platonic solids as archetypes when constructing the solar system." (Voelkel 2001, 4).

[14]"Kepler's view of the relationship between appearance and reality is predicated on the idea that material substance is obedient to a divine rational plan and, indeed, was created to instantiate the archetypes. So while there is a distinction between appearance and reality, there is also a systematic relationship between them that can be explored and exploited by observing appearances in tandem with considering the causes a priori." (Martens 2000, 68).

We must keep in mind that contemporary astronomy, just like any other system, is rooted in its culture and is a product of many assumptions about the way the world works, what it represents and means, and its ultimate source.

References

Aristotle, *De Caelo Nicomachean Ethics* Plato, *Timaeus*

Bertemes, Francois, Peter Biehl, Andreas Northe, and Olaf Schroder, 2004, "Die neolithische Kreisgrabenanlage von Goseck, Landkreis Weißenfels" in *Archäologie in Sachsen-Anhalt/ N.F.* 2 pp. 137–145

Chippindale, Christopher, 2004, *Stonehenge Complete, Third Edition* (New York: Thames and Hudson)

Dicks, D.R., 1970, *Early Greek Astronomy, to Aristotle* (Ithaca: Cornell University Press)

Gingerich, Owen, 1975, "The Basic Astronomy of Stonehenge", in Brecher, Kenneth and Michael Feirtag (eds.), *Astronomy of the Ancients* (Cambridge MA: MIT Press)

Heidarzadeh, Tofigh, 2008, *A History of Physical Theories of Comets, from Aristotle to Whipple* (New York: Springer)

Kukkonen, Taneli, 2002, "Infinite Power and Plenitude: Two Traditions on the Necessity of the Eternal", in Inglis, John (ed), *Medieval Philosophy and the Classical Tradition: In Islam, Judaism and Christianity* (London: Routledge)

Martens, Rhonda, 2000, *Kepler's Philosophy and the New Astronomy* (Princeton: Princeton University Press)

Murphy, Anthony and Richard Moore, 2006, *Island of the Setting Sun: In Search of Ireland's Ancient Astronomers* (Dublin: Liffey Press)

Pankenier, David, 2013, *Astrology and Cosmology in Early China: Conforming Earth to Heaven* (Cambridge: Cambridge University Press)

Pearson, Michael, 2012, *Stonehenge: Exploring the Greatest Stone Age Mystery* (New York: Simon and Schuster)

Parker Pearson, M. G., Chamberlain, A., Jay, M., Marshall, P., Pollard, J., Richards, C., Thomas, J., Tilley, C., Welham, K., 2009, "Who was buried at Stonehenge?" *Antiquity: a quarterly review of archaeology* 83, 23–29

Sun, Xiaochun and Jacob Kistemaker, 1997, *The Chinese Sky During the Han: Constellating Stars and Society* (Leiden: Brill)

Voelkel, James, 2001, *The Composition of Kepler's Astronomia Nova* (Princeton: Princeton University Press)

Chapter 9
Contemporary Astronomical Thought

This brings us to the present day. The main focus of this book has been ancient astronomy, with a particular focus on Non-Western astronomy. Contemporary astronomical thought can be seen through the lenses of this. In a sense, I'm inverting the standard history of astronomy one sees, with a small section on ancient and Non-Western astronomy, and an increasing amount as we get closer to contemporary thought. This is not an illegitimate picture of astronomy, but it's one that takes contemporary astronomy as the pinnacle of human thought about the cosmos, to which past history has led. Those systems unconnected (or at least not obviously connected) to contemporary astronomy are thus neglected. This is an akin to an astronomical version of "whig history", in which there is a narrative of continual development to more enlightened, sophisticated, or true understandings of the heavens as humanity moved into the future. It the kind of story we in the contemporary world like to tell. It makes us proud of our own accomplishments. But, as with anything else, a sense of our own skill and importance need not require denigrating or denying the accomplishments of others. Only the most fragile and minute feel threatened by recognizing the abilities, talents, discoveries, and understanding of others, instead of learning from them.

The role of this book is not, then, to give an account of how contemporary scientific astronomy came to be as it is. Instead, we have looked at various ways of understanding the sky in the ancient world and contemporary world, in order to better understand their unique features, as ell as to learn from them. Thus, in this final chapter, I look briefly at important developments in observational and theoretical astronomy in the contemporary world, with a view toward explaining the concerns and philosophical assumptions grounding our understanding, and anticipating how this might develop in the future.

Today, our understanding of the motions of the solar system bodies is basically that of Kepler, of course with additional improvements and variations. But the basic laws he discovered are still accepted as expressing the motions of the planets and the earth. These laws are still taught in classes in astrophysics (I learned it in my

A. McLeod, *Astronomy in the Ancient World*,
Historical & Cultural Astronomy, DOI 10.1007/978-3-319-23600-1_9

own first class in this area years ago). These laws are still simple, elegant, and powerful enough to awe a person who first encounters them, and I still remember the day I learned these simple principles that explained in such a succinct and straightforward way motions that had taxed and vexed astronomers in the West for thousands of years before Kepler's time. No doubt, it was the innovations in observation and the possibility of very precise measurements to accurately plot coordinates of celestial objects that made this possible. Today, our methods of observation are many times better than those of Tycho Brahe, yet his observations were enough to give us the understanding of celestial motions we retain to this day. And our ability to observe objects far beyond the solar system as well as theories that account for the behavior of these objects has vastly expanded our view of the size and complexity of the cosmos beyond the view of Kepler and his contemporaries. Although the motions of objects in the solar system are important, cosmology today is not understood in terms of the planetary or solar level, but on a much larger scale. Still, Kepler's understanding of the basic motions of the planets and solat system objects in general is one we still hold today, and formed the basis of contemporary Western thinking about celestial motions to follow, through the influence of Isaac Newton. Thus, in discussing at least a part of our contemporary views on the structure of the cosmos, it is necessary at the same time to discuss Kepler and his laws. It is for this reason that some (including myself) mark Kepler's work as signaling the beginning of the modern scientific era with respect to astronomy.

The scientific attitude itself is one that must be understood critically. It is not as if the contemporary concerns, methodologies, and ways of understanding the world were born fully formed from the head of 17th century Europe. Nor is it the case that we can take our current understanding as *definitive* or final in a way that others have not been. Our own ways of undertstanding the cosmos (like any other feature of our lives) are reflections of our more general concerns, assumptions, cultural patterns, and ways of interacting with the world. We fool ourselves if we take the contemporary scientific pursuit as completely objective and predicated by the "view from nowhere". Our contemporary understanding of scientific method, progress, and understanding itself must also be investigated critically, and not simply taken as holy writ. Indeed, it is not altogether clear that contemporary scientific method is even one single thing at all, or that it works as cleanly as we present it and sometimes believe. In this final section, then, I will offer a brief account of the story of the cosmos from the perspective of contemporary professional scientific understanding, but to take this as a revealing of "the true story" after all of the mythological, superstitious, uniformed, or otherwise untrue stories of the ancients would be unwise and equally unjustified. There is no independent criterion to determine whether the view of the stars stripped of certain aspects of their human significance is a better or truer one than others. The human cannot be fully stripped of it, no matter how hard one tries—the human observer is always part of the observed. We understand the cosmos through our own minds, we sense it (all empirical science is ultimately based on the senses) through our bodies, all of which are contingent and unique, and *located*. We are not objective and perspectiveless instruments. No instrument is. Every tool has something it does, and things it cannot do. If we are like tools, we are *changing* tools, whose function can be determined in part by things like culture, philosophical assumptions, concerns, etc. The contemporary scientific understanding of the sky becomes interesting as a comparative tool against the background of the ancient views when we take it as another way or an alternative way of understanding, one with its own background assumptions and concerns, instead of as the "right" understanding which supersedes all of the others. There is no scientific method that can tell us that the view of the sun as a collection of gas so densely packed that it creates nuclear fusion is any more accurate or true than the view of the sun as representative of life, continuity, and the ultimate human identity with nature that we plausibly find in people like the builders of Newgrange.

Contemporary cosmology is connected closely to observational astronomy in ways we did not always see in early cultures. Part of the reason for this is the power and extent of observation possible with contemporary technology. Not only can we observe visible objects at much greater distances with much higher detail than ever before, but we can also observe in different ranges of the electromagnetic spectrum beyond that of visible light, such as the ultraviolet, infrared, radio, and microwave. These forms of observation, combined with contemporary theories, have led to major discoveries and have radically shaped our understanding of the sky. In order to tell the story of contemporary cosmology, I will leap forward a bit and first discuss some of the technologies which continue to make understanding in this area possible.

The development of telescopes in the modern period took off with the innovations of William Herschel, a musician turned self-taught astronomer, in the late 18th century. Herschel's meticulous construction and the precision of his instruments quickly made them the best telescopes available, better than anything possessed even by "establishment" astronomers, who soon turned to him for their instruments, especially after his discovery of the planet Uranus (probably the event for which he is best known today). Makers of telescopes began to build larger instruments than ever before, including Herschel, who constructed a massive 49.5 in telescope, with a 40 ft focal length (the largest telescope in the world at the time and for almost half a century to come). The ability of the (reflector) telescope is determined primarily by the diameter of the primary mirror. The larger the mirror, the more light from the sky that can be collected. Contrary to popular belief, the primary mirror diameter has nothing to do with *magnification*. It does not make the images of stars or other celestial objects viewed through a telescope any larger. However, it does make them more *resolvable*, and thus we can maintain good resolution using higher magnification eyepieces than we can with telescopes of a smaller diameter. For example, a magnification level on my 8 inch telescope that might render a planet as an unintelligible blob will give a beautiful and sharp rendering of the same planet on a 40 in telescope (though it won't be in color—more on that below).

Telescope development continued through a phase of making larger and larger aperture telescopes, until inevitably the problem became one of dealing with the massive size of these large scopes. Even Herschel did not use his "40 ft" telescope much for observing, as it was unwieldy. It's not easy to move such a gigantic

(a) (b)

Fig. 9.1 **a, b** Two telescopes by the legendary William Herschel. On the *left* is a drawing of the infamous 40 ft telescope (with an indication on the *lower right* of how the telescope was moved). On the *right* is one of Herschel's telescopes in the collection of the Adler Planetarium, Chicago, IL. Photo by the author

structure, even though Herschel certainly did the best he could given what was available at his time. The telescope had an altitude- azimuth mounting (as all of Herschel's telescopes did), and thus had to be moved in two directions—around the circle of the horizon, and directly up and down toward the zenith. To do this, Herschel created a rope and pulley system that had to be operated from cranks on the telescope platform that, even with the reduced force from the pulleys, cannot have been easy to operate, and it required two people to do it (independently of the observer) (See Fig. 9.1).

There are a couple of additional problems with large telescopes. For reflectors, the mirrors in large telescopes needed to be polished and cleaned more often than those of smaller telescopes, in part because of the quality of the mirrors they were working with, which also made the resolution at these much greater diameters minimally better (or not better at all) than in smaller telescopes. It was not until new techniques and technologies came along in the 19th and 20th centuries that it became feasible and useful to build much larger reflecting telescopes, such as George Hale's 60 in telescope at Mount Wilson Observatory (built in 1908), and the 200 in telescope at Palomar Observatory relatively nearby (built in 1948), named after Hale. The problem with massive size in refracting telescopes is more straightforward. As we know, the more mass something has the greater effect gravity will have on it. Since refractors require *lenses* rather than mirrors, there can be nothing behind these lenses holding them into place, as is possible with a mirror, and they must then be held into place by their edges, allowing for light to travel through. While relatively small lenses will present no problem, as we build larger and larger lenses, the stress on the center of the glass will be greater. Since there can be nothing to hold this part of the glass in place (while the edges are held into place), this will lead to warping of the lens, which will drastically affect the quality of the telescope. It will not take too much warping for a scope to become completely unusable. And it is not as easy a task to switch out a lens in a refractor as it is to change a mirror in a reflector.

To this day, the largest telescopes are reflectors, for the reasons mentioned above. As the technology to maintain polish and shape of mirrors improved, the aperture of reflecting telescopes grew with it, leading to higher and higher resolvability. Today, the largest refracting telescope is a mere 40 in (at Yerkes Observatory in Wisconsin), and was built in 1897. The largest reflector, on the other hand, is the Gran Telescopia Canarias in La Palma, Spain, which boasts a 34.2 ft mirror (410.4 in), and opened in 2007. In visible light astronomy, reflectors clearly play the larger role today.

There is another difficulty that limited all telescopes, reflector and refractor alike, until relatively recently in human history—one that was harder to solve than the problems with mirrors and lenses. One key limitation on our ability to resolve celestial objects is the natural barrier we all have to look through: the earth's atmosphere. The layers of gas that comprise our atmosphere act the same way any other medium through which we can see acts—distorting the light that passes through it. While we generally don't use air or water or other similar materials to lens light for observations (in part because of their volatility), the principles behind the use of lenses for eyeglasses, refracting telescopes, or microscopes is in large part the same as those explaining the phenomena we notice when looking through water or air. At its root, each of these cases involves changing the way light waves move through space and are thus collected by the observer in the eye (or other instruments). At the extreme end, even the effects of gravity itself can create distortions of light waves (gravitational lensing, for example, was one of the key discoveries further confirming Einstein's general theory of relativity, which predicted this phenomenon). When light travels through a medium such as water or air, it is refracted in the same way it is when traveling through a lens. The difference is that the lens is made so as to control the direction of refraction in a way we can use it to resolve a diffuse amount of light into a smaller area (this is the same reason it is sometimes possible to create a fire by using a magnifying glass on the backdrop of the sun—it takes the sun's already very strong light captured on the surface of the glass and condenses it down to an even smaller point, thus increasing the heat at the focal point). We can see this refraction when we look at a straight object, such as a pencil, placed into water. The object looks bent when in the water, due to the refraction of light by the medium of the water, making the part of the pencil in the water appear in a different location. The same process happens when we view objects through the atmosphere. The atmosphere not only skews the apparent locations of celestial objects, but also can make them harder to observe, due to the volatility of the atmosphere. Fluctuations and layering in the atmosphere cause the light emitted from distant stars and other objects to scatter. This is the reason for the phenomenon we know as the "twinkling" of stars. Clearly, the atmosphere itself is part of the problem when it comes to observing. If we could get outside of it, telescopes would be much more efficient.

In relatively recent years, there have become two main ways to remove or limit the obstacle presented by the atmosphere. The first one, even though difficult, is somewhat more straightforward. Looking through the atmosphere is no longer a problem if one can get *outside* of the atmosphere. Thus, a number of telescopes (within the last half century or so) have been carried into space. Probably the most famous of these is the visible-light telescope that has given us so many magnificent images over the years since its launch in 1990, the Hubble Space Telescope. The second way of getting around the obstacle of the atmosphere is the even more recently developed system of adaptive optics. The principle behind adaptive optics is simple enough—since the light from stars viewed through the atmosphere is distorted by the atmosphere, then it should be possible to compensate for this by creating distortions in the reflecting mirror canceling this atmospheric distortion. And this is essentially how adaptive optics works.

There are limitations to what we can observe through visible-light astronomy. Just as we notice on earth not all light (electromagnetic radiation) is in the visible spectrum, the same is the case elsewhere in the universe. If we can observe the skies with instruments that receive different wavelengths of light beyond the visible, in different areas of the spectrum, then we will be able to see different objects or phenomena radiating light in non-visible wavelengths.

And at this point William Herschel returns to our story. In the winter of 1800, Herschel was in the middle of a series of solar observations. In order to safely observe the sun, one needs to look through a filter that reduces the magnitude of the light of the sun enough for it to be safely viewed, and enough for one to see detail on the sun, such as sunspots, flares, etc. Herschel was testing a number of filters on a telescope to determine which made for best observing of the sun. He found that filters of different colors seemed to generate different amounts of heat, and wondered if this might signal differences in the sun's radiation at different wavelengths. He devised a simple experiment, using the sun's light, filtered through a prism. Noticing that a red filter caused the greatest heat, Herschel held a thermometer at the red end of the visible spectrum of the solar light filtered through the prism, and then just past the red end, and found that the temperature was higher past the red end. This led him to the view that the sun must be radiating light in some invisible wavelength, which he called "calorific rays", but we know today as *infrared* radiation. Herschel, the prodigious self-taught astronomer and telescope maker, discoverer of Uranus, had discovered the existence of light beyond the visible spectrum. Two infrared telescopes today are named after him, in honor of his discovery—the optical "near-infrared" William Herschel Telescope in Spain, and the true infrared Herschel Space Observatory, recently decommissioned in 2013, when it became inoperable after a lengthy mission of almost 4 years (See Fig. 9.2).

Today, astronomy in non-visible wavelengths dominates observation, with visible-light astronomy relegated to a relatively small corner. Part of the reason for this is that many more discoveries are made observing in non-visible wavelengths,

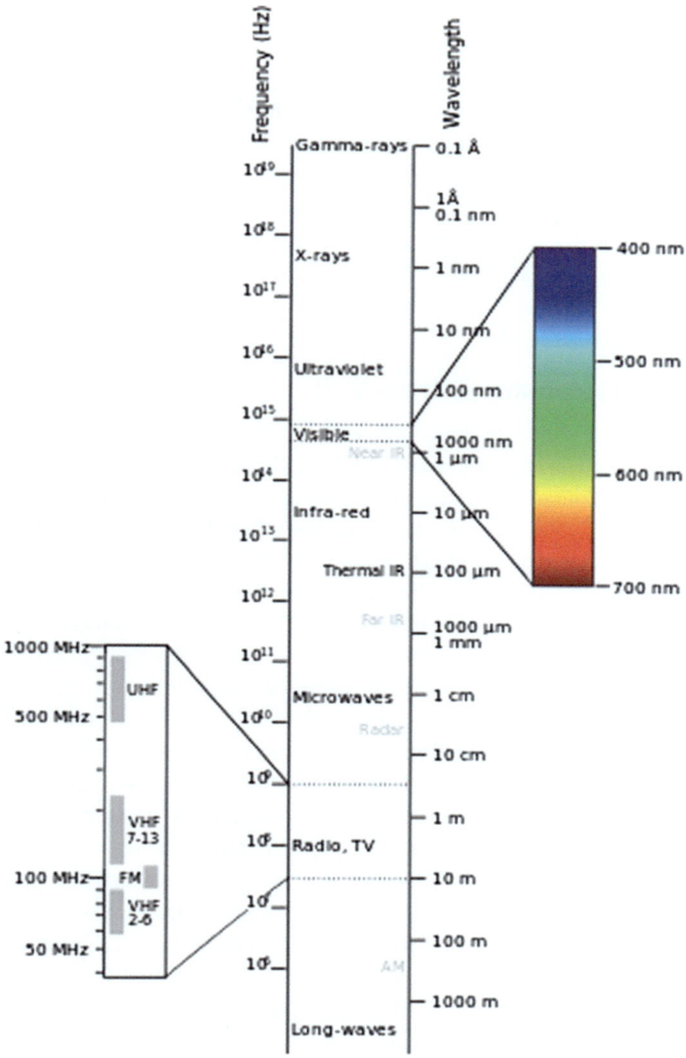

Fig. 9.2 The electromagnetic spectrum. A chart showing the wavelength ranges and relationships of the different parts of the spectrum. Visible light occupies a relatively narrow range, from 400 to 700 nm. Much of contemporary astronomy relies on observations made of radiation in wavelengths below and above that of visible light (Photo credit: Victor Blacus, Creative Commons 3.0 https://en.wikipedia.org/wiki/File:Electromagnetic-Spectrum.svg)

due to the nature of observation. When we look in ways we have never seen before, we will necessarily find many things that we have never seen, which will give us new insights into how astronomical objects operate. We have seen the pinpoint visible light of distant stars with our eyes, for example, since the beginning of humanity. But we have never seen the non-visible radiation of stars, or understood it at all until relatively recently in human history. Visible-light astronomy has a far longer history than non-visible astronomy, and the discoveries made observing in different wavelengths of the electromagnetic spectrum have been enormous and plentiful, transforming our understanding of celestial phenomena.

Astronomy today is done in numerous areas of the spectrum beyond the visible. Infrared, radio, ultraviolet, x-ray and gamma ray astronomy are all major areas of research. Discoveries continue to be made using all of these. There are many examples of such discoveries, but I will focus on just a couple of them here, as examples.

Ultraviolet Astronomy-Mira's Tail

In 2003, Professor Christopher Martin of the California Institute of Technology and his colleagues at Caltech, the nearby NASA Jet Propulsion Laboratory, and other academic and private institutions launched the Galaxy Evolution Explorer (GALEX) space telescope. GALEX, which observed in the ultraviolet, was put into a low earth orbit, and Martin and his colleagues were able to make observations of various parts of the sky in the ultraviolet. One of the advantages the GALEX telescope had was that it was able to observe without much interference from sunlight, which pervades our sky even during the nighttime. As we saw above in consideration of Herschel's discovery of infrared light, the sun radiates most strongly in the red end of the visible spectrum and the infrared, and less so in higher frequencies of the electromagnetic spectrum such as the ultraviolet. It does radiate in the ultraviolet (as we know from the need for sunblock when it's sunny outside, as ultraviolet A can cause skin cancer), but it radiates far less in this frequency. Thus, in the ultraviolet, the sun is far less "bright", and this makes for better observing conditions. Just as in optical astronomy, strong visible light makes observation difficult. Bright light drowns out dimmer light, and the light we receive from the distant stars is much dimmer than most light emitted by sources closer to home, such as lightbulbs. This is the case with the sun and moon as well. In the daytime, the brightness of the sun eliminates any view of most celestial objects other than the moon. This clashing of light can only happen when the light received is in the same frequency. Thus the bright visible light of the apparent magnitude of the sun drowns out the dim visible light of the stars, of far less apparent magnitude (although we have to say their apparent magnitude is *higher* rather than lower, as the scale of magnitude is measured from lower numbers at brighter objects—the sun's apparent magnitude is −27 while that of the bright summer star Vega is around 0, and that of Polaris (the North Star) is +2. The initial reason for this was that stars were ranked into *classes* of brightness—the brightest stars being in the first class, the next in the second, and so forth). In the ultraviolet, on the other hand (as with other frequencies), visible light will not drown out the light. Thus, if we are observing in the ultraviolet and someone turns on a light right near us which does not radiate at all in the ultraviolet, we will not even notice it. We won't see it. This is the same reason that we don't see radiation in other wavelengths. The heat signatures picked up on infrared devices are invisible to our eyes, which receive only visible light.

In their survey of the sky, the GALEX team discovered a number of interesting things previously unknown to observers.[1] One of the strangest and most interesting was their discovery about the star Mira. Mira is a binary star system in the constellation Cetus, comprised of the variable star Mira A and the smaller Mira B orbiting it. Mira has an interesting history. It has been known since ancient times, but only recognized as a binary in the early 20th century. In addition, its variable apparent magnitude makes it unique, in that at its brightest it is easily visible, while it is completely invisible to our eyes at its lowest magnitude. It swings between magnitude 2 at brightest and magnitude 10 at dimmest. Magnitude 2 is easy to see with the naked eye alone given a dark sky (though still impossible for most of us to see who live in developed areas, due to rampant light pollution), while Magnitude 10 is far beyond the limiting magnitude for the naked eye in even the darkest and clearest skies. Mira turns out to be the brightest variable that is invisible for part of its period of variability.

The GALEX team, during their survey of the sky, noticed on one of their images a faint nebulosity toward one of the corners. It seemed to be a kind of gaseous halo surrounding a star, which turned out to be the variable Mira. They moved the UV telescope to center Mira and get a better view, and found something even more fascinating. This image that appeared at first to be a halo extended in a long trail all the way through the field of vision and beyond! It was as if the gas halo surrounding Mira had been pulled out and extended on one side of it, just like the tail of a comet. However this nebulosity around Mira was far longer than a comet's tail. When the GALEX team traced the entire "tail", it turned out to extend a full 2 degrees of arc—a full four times longer as a stretch of the sky from the earth's perspective than that of the moon. Of course, the amount of arc of the sky an object takes up relative to our perspective is a function of both its size and its distance. The moon is almost directly on top of us from an astronomical perspective, only 200,000 mi or so away. Mira, on the other hand, is about 300 light years away. Thus, the actual (rather than apparent) length of the tail is 13 light years—about three times longer than the distance between earth and the nearest (extrasolar) star. Even with such a massive size, the reasons we are unable to see this "tail" are its dimness and the fact that it is not radiated in visible light, but in the ultraviolet. This is why the UV telescope of GALEX was necessary for its discovery.

What appears to be a comet-like tail turns out be the result of an unusual phenomenon. The bow shock (the place in which the interstellar medium meets the magnetic field of an object) combined with the solar wind created by the high speed of the system (130 km per second) strips mass from the system that creates the long "tail" (See Fig. 9.3).[2]

[1]The GALEX team describe their findings in Martin et al. 2007.

[2]Martin et al. 2007, Wareing 2008.

Fig. 9.3 The binary star Mira, and its massive "tail", extending 13 light years. This tail does not radiate in the range of visible light, and thus was only discovered when observing the star in the ultraviolet. Astronomical observation in non-visible wavelengths continues to transform our understanding of the cosmos (Photo credit: NASA)

Cosmic Background Radiation

A major discovery for contemporary cosmology was made in 1965 through observation with a radiometer, which is a modified radio device designed to pick up the power of a particular frequency from a radio telescope. Astronomers Arno Penzias and Robert Wilson were the first to detect and measure the *cosmic background radiation* a radiometer they had built for a different purpose altogether. This cosmic background radiation they discovered is in the microwave wavelength, and though Penzias and Wilson were the first to empirically demonstrate its presence, its existence had been theorized for years by the time they made their discovery. It was first predicted in 1948 by Ralph Alpher and Robert Herman, adherents of the "Big Bang" theory of cosmology.[3] If the theory were correct, then there should be a remnant radiation from the Big Bang permeating the expanding universe. The radiation discovered by Penzias and Wilson was consistent with the predictions for the model, and thus played a major role in helping the Big Bang theory overcome its competitors, particularly the "steady state" model. The Big Bang model, as we all know, is the accepted theory of the origin of the universe today. Penzias and Wilson were awarded the Nobel Prize in 1978 for their work.

The discovery of the cosmic background radiation and corroboration of the Big Bang theory is just one of the myriad discoveries made possible by the incredible predictive power of modern scientific astronomy. Astronomers continue to use the methods of modern science to understand our universe. In modern years, and on the frontiers of astronomy, theoreticians and experimental scientists have uncovered long hidden facts about black holes, dark matter, planets of solar systems outside of our own (exoplanets), the planets in our own neighborhood, the formation of galaxies, and the origins of the universe. The future for astronomy is certainly bright, but it is also important that we understand the ultimately contingent nature of contemporary scientific astronomy, and its explanatory limits.

[3] A phrase actually coined by Fred Hoyle, an opponent of the theory and adherent of the "steady state" model (Mitton 2011, 128).

The Philosophy of Two-World Naturalism and the Future of Astronomy

Contemporary astronomy, as much as any of the systems covered in this book, is based on underlying worldviews and philosophical assumptions. Assumptions, by their very nature, lack solid grounding and justification. They are based on intuitions. Of course, intuitions are not manifestly bad things or things to be avoided, as they are sometimes all we have, and we can be more or less careful in the way we employ our intuitions. Knowing the nature and limitation of intuitions will allow us a greater respect for an appreciation for systems based on different assumptions, grounded in different intuitions. We cannot deny the predictive power of the modern scientific worldview and its picture of the cosmos. This predictive power has enabled us to build all of the machinery of the contemporary world, and to travel to new worlds such as the moon and the other planets of the solar system. We have landed humans on the moon, landed rovers on the moon that continue to collect valuable information, and have visited every known planet of the solar system with the machines we have created. However, the fact that modern science does have such material predictive power does not in itself show us that the results of modern science alone give us an "objectively true" account of the world, any more than the behavioral power of Chinese astronomy (in its ability to legitimize political entities and organize disparate elements of empire) show us its sole and objective truth, or the uniqueness and complexity of Maya calendars and mathematics shows the same. In addition, our own *version* of science is not the only possible version. The ancient Maya, Cahokians, Chinese, Indians, and others all had scientific understandings of their world in a broad sense of the term. While their thought certainly didn't share all of the features of contemporary scientific thought, they understood their world in ways that mirror the ways contemporary people understand it, including depending on empirical data, hypothesis, and experimentation. While science is not the sole creation of the West, we must still look beyond only science if we want to truly understand our universe and our place in it, and beyond only contemporary scientific astronomy.

While recognizing the power and usefulness of contemporary astronomical thought (about astronomy or anything else), and the vast amount of knowledge we have gained from it, we thus ought to retain intellectual humility. We should recognize the ways in which the very different astronomical systems of world history, based on different philosophical, religious, and other assumptions, may teach us important things about the universe in which we live that we do not and *cannot* discover using the methodology of contemporary scientific astronomy. Contemporary science and the astronomy based in it are incredibly powerful, and surely among the greatest achievements of humankind. The discoveries, methodologies, and ways of thinking of a number of people in different times and places throughout the world have contributed to the development of this contemporary ideology. But they have also in key ways *challenged* it, and we should take those challenges as seriously as we take the contributions. We should measure the value of a system not in terms of how closely it mirrors our own understandings of the world (this is to assume that we have all the answers), but in terms of how it might contribute to a broader and more complete understanding of our world.

In addition, we have to be careful to avoid a view that holds that astronomy (and any other human pursuit) progresses and improves as we move into the future. Such a view naturally rejects the astronomical thought of the past as outdated, primitive, or simply untrue. This kind of view of the " evolution" and development of human understanding of the world, the scientific counterpart to whig history, is based at least in part on an improper understanding of the phenomenon of evolution—one which the Social Darwinist movement also made, albeit in a different way. Creatures evolve in relationship with their environment, and evolution happens as adaptation to environments. Even though the word 'evolution' in the popular consciousness now has connotations of value, positive development, or growth into something superior, this is not at all what evolution is. Given the vagaries of environments, evolution could result in the diminishing or elimination of certain abilities or capacities, just as it could result in the gain or improvement of them. The assumption that creatures somehow get *better* or more able through evolution is simply false. Evolution is adaptation to environment.

The idea that contemporary understandings of the cosmos are necessarily a progress and development beyond earlier understandings, then, is enormously problematic. A large part of the problem is that there is simply no reason to believe it. There is no evidence that human understanding has a straight (or even crooked) line of progress as we move into the future. Indeed, there are many things the ancients understood that we no longer do—key features of human life that we have lost, and that perhaps contribute to the social crises of the modern world.[4] The human connection to the sky is something we have for the most part lost. We obscure our sky with light pollution, and have little access to the stars. We have become disconnected and in some sense alienated from the rest of nature in part through our own attempts to shape nature. We no longer feel the connection to the processes of nature represented by structures such as Newgrange, the Mesoamerican pyramids, the Forbidden City, or Jai Singh's observatories. This is likely part of the reason so many people compete in the lottery to get inside the chamber at Newgrange on the winter solstice every year. We are trying to reconnect, to bring back something we've lost. We no longer have a ritual sense of the processes of nature, through the operation and maintenance of observatories like those of Jai Singh. We no longer have the concern for the sky as a meaningful and important part of human life that can tell us something about what we are, our purpose, and our ultimate fate. And for these reasons, astronomy falters and fades from our collective consciousness. It becomes the purview of specialists, scientists, and a dwindling and aging number of amateur enthusiasts.

[4]Sociologist Max Weber (1864–1920) was a notable proponent of such a view. "In diagnosing that societal rationalization processes would inevitably result in a loss of freedom (*Freiheitsverlust*), whereas Western cultural rationalism would lead to a loss of meaning (*Sinnverlust*), Weber provided a highly ambivalent evaluation of the significance of modernity in the West." (Benhabib 1981, 115).

All of that said, we must be careful to avoid the pitfall of idealizing the past and ancient systems as well. Just as contemporary understandings inevitably have flaws, blind spots, and other problems, ancient understandings did as well. Some of these mistakes we can now see, with the perspective of distance, in a way the people involved could not. Some of the problems we still may not see, just as we will necessarily miss many of our own. Some of those people of the past may have been able to point out additional problems with our own approach, and people in the future will likely be able to do this as well. The advantage of distance should not be underestimated. As with history, the fullest understanding comes when one is not in the middle of an event. When we have time to reflect and look back. Of course, we always must appraise the ideas of the past based on our own views and assumptions, but we can do this with intellectual humility when we recognize that our own assumptions (mechanistic naturalism, non-dualism, etc.) are themselves merely assumptions and not something more. We can often forget this and instead take our foundational beliefs as somehow necessary, irrefutable, or justified, empirically or otherwise. We cannot empirically or otherwise demonstrate that materialism is true and idealism, for example, is not. In order to truly understand and appreciate the astronomical understanding of the past (or any other aspect of early society), we have to recognize the conditional and limited nature of our own understanding of the world. If we simply accept (and this can never be based on evidence) the ultimate truth and superiority of our own background assumptions and the systems of thought based in them, we will constantly misunderstand, misconstrue, and fail to learn from other cultures of the past and present alike, and our own culture will thereby grow stagnant.

I hope this book has helped to demonstrate, at least in some small way, the power and importance of systems of astronomical thought around the world, ancient as well as modern and contemporary, in every section of the world. There are many more systems to discover and learn from. There are many interesting systems I have not included in this book, such as those of the ancient Babylonians, Egyptians, Polynesian islanders, various peoples of the Americas, North and South, not covered here, people of the vast continent of Africa, and many others. Perhaps one day we will come to more widely recognize the significance of these systems.

References

Benhabib, Seyla, 1981, "Modernity and the Aporias of Critical Theory", *Telos*, 49, pp. 39–59

Hoskin, Michael, 2011, *Discoverers of the Universe: William and Caroline Herschel* (Princeton: Princeton University Press)

Martin, D. Christopher, Mark Seibert, James Neill, David Schiminovich, Karl Forster, R. Michael Rich, Barry Welsh, Barry Madore, Jonathan Wheatley, Patrick Morrissey, and Tom Barlow, 2007, "A Turbulent Wake as a Tracer of 30,000 Years of Mira's Mass Loss History" *Nature*, 448, pp. 780–783.

Mitton, Simon, 2011, *Fred Hoyle: A Life in Science* (Cambridge: Cambridge University Press)

Wareing, Christopher, 2008, "Wonderful Mira", *Philosophical Transactions of the Royal Society A,* 366 (1884), pp. 4429–4440.

Index

A. McLeod, *Astronomy in the Ancient World*,
Historical & Cultural Astronomy, DOI 10.1007/978-3-319-23600-1

MIX
Papier aus verantwortungsvollen Quellen
Paper from responsible sources
FSC® C105338

If you have any concerns about our products,
you can contact us on
ProductSafety@springernature.com

In case Publisher is established outside the EU,
the EU authorized representative is:
Springer Nature Customer Service Center GmbH
Europaplatz 3, 69115 Heidelberg, Germany

Printed by Libri Plureos GmbH
in Hamburg, Germany